WITHDRAWN

Finite Element Analysis of Shells of Revolution

Surveys in Structural Engineering and Structural Mechanics

Main Editor
W. F. Chen, Purdue University

Advisory Editors

J. H. Argyris, University of Stuttgart
Z. P. Bazant, Northwestern University, Evanston
T. B. Belytschko, Northwestern University, Evanston
P. G. Bergan, University of Trondheim
D. C. Drucker, University of Florida, Gainesville
T. V. Galambos, University of Minnesota
M. P. Gaus, National Science Foundation, Washington, DC
K. H. Gerstle, University of Colorado
C. Massonnet, University of Liège
G. Mehlorn, University of Kassel
Z. Mróz, Institute of Fundamental Technological Research, Warsaw
F. Nishino, University of Tokyo
E. P. Popov, University of California, Berkeley
E. C. Ting, Purdue University
N. S. Trahair, University of Sydney
R. Wang, Peking University
J. Witteveen, Institute of Building Materials and Building Structures, Delft
J. T. P. Yao, Purdue University
O. C. Zienkiewicz, University College of Swansea

Finite element analysis of shells of revolution

Phillip L. Gould
Department of Civil Engineering
Washington University, St. Louis

Pitman Advanced Publishing Program
Boston · London · Melbourne

PITMAN PUBLISHING INC
1020 Plain Street, Marshfield, Massachusetts 02050

PITMAN PUBLISHING LTD
128 Long Acre, London WC2E 9AN

Associated Companies
Pitman Publishing Pty Ltd, Melbourne
Pitman Publishing New Zealand Ltd, Wellington
Copp Clark Pitman, Toronto

© Phillip L. Gould 1985

First published in Great Britain 1985

Library of Congress Cataloging in Publication Data

Gould, Phillip L.
  Finite element analysis of shells of revolution.
  (Surveys in structural engineering and structural mechanics; 4)
  Bibliography: p.
  Includes index.
  1. Shells (Engineering) 2. Finite element method.
  I. Title. II. Series.
  TA660.S5G645 1985   624.1′7762   84-26600
  ISBN 0-273-08654-5

British Library Cataloguing in Publication Data

Gould, Phillip L.
  Finite element analysis of shells of revolution.
  —(Surveys in structural engineering and structural mechanics; 4)
  1. Shells (Engineering) 2. Finite element method
  I. Title   II. Series
  624.1′7762   TA660.S5
  ISBN 0-273-08654-5

All rights reserved. No part of this publication may be reproduced, stored in a retrieval system, or transmitted, in any form or by any means, electronic, mechanical, photocopying, recording and/or otherwise, without the prior written permission of the publishers. This book may not be lent, resold, hired out or otherwise disposed of by way of trade in any form of binding or cover other than that in which it is published, without the prior consent of the publishers. This book is sold subject to the Standard Conditions of Sale of Net Books and may not be resold in the UK below the net price.

Filmset and printed in Northern Ireland by The Universities Press (Belfast) Ltd., and bound at the Pitman Press, Bath, Avon.

To the many research students who contributed so much to this book.

Contents

Preface xiii

1 **Introduction** 1
  1.1 Description of Structural Form 1
  1.2 Scope of Treatment 1
  1.3 Historical Review 3
  1.4 Notation for Matrices 4

  References 4

2 **Fundamentals** 7
  2.1 Surface Geometry 7
  2.2 Ring Element 8
  2.3 Kinematic and Static Variables 10
    2.3.1 Definitions 10
    2.3.2 Fourier Series Representation 12
  2.4 Kinematic Laws 14
    2.4.1 Transverse Shearing Strains 14
    2.4.2 Linear Form 14
    2.4.3 Nonlinear Form 16
  2.5 Constitutive Laws 17
  2.6 Boundary Conditions 19
  2.7 Variational Principles 20
    2.7.1 Hamilton's Principle 20
    2.7.2 Principle of Minimum Stationary Potential Energy 22
    2.7.3 Reissner's Principle 22
    2.7.4 Solution of Variational Problems 24

  References 24

3 **Static Analysis** 26
  3.1 Representation of Dependent Variables 26
    3.1.1 Comparison Functions 26

Contents

   3.1.2 Approximations at Element Level   28
3.2 Linear Displacement Formulation   29
   3.2.1 Element Equilibrium Equation   29
   3.2.2 Condensed Equilibrium Equation   31
   3.2.3 Global Equilibrium Equations   32
      3.2.3.1 Assembly   32
      3.2.3.2 Boundary Conditions   33
      3.2.3.3 Solution   33
      3.2.3.4 Strains and Stress Resultants   34
      3.2.3.5 Rigid Body Displacements   35
3.3 Extensions to Basic Element   36
   3.3.1 General   36
   3.3.2 Stiffened Shells   36
      3.3.2.1 Classification   36
      3.3.2.2 Circumferential Stiffeners   36
      3.3.2.3 Meridional Stiffeners   38
   3.3.3 Compound Shells   38
   3.3.4 Branched Shells   40
   3.3.5 Open-Type Elements   41
      3.3.5.1 General   41
      3.3.5.2 Member Displacement Field   42
      3.3.5.3 Member Stiffness Matrix   43
      3.3.5.4 Coordinate Transformations   46
      3.3.5.5 Reduced Member Stiffness Matrix   48
      3.3.5.6 Effective Stiffness Matrix   49
      3.3.5.7 Consistent Load Vectors   51
   3.3.6 Local Effects of Columns on Shell   52
3.4 Mixed Formulation   55
3.5 Convergence and Discretization Criteria   56
   3.5.1 Convergence   56
   3.5.2 Discretization Criteria   56
3.6 Case Studies   58
   3.6.1 Cylindrical Shell under Edge Loading   58
   3.6.2 Cylindrical Shell under Hydrostatic Loading   59
   3.6.3 Parabolic Shell under Antisymmetrical Loading   64
   3.6.4 Hyperboloidal Shell under Static Wind Loading   66
   3.6.5 Cylindrical Shell under Thermal Loading   66
   3.6.6 Cylindrical Shell with Torospherical Head   66

References   75

4 Dynamic Analysis   77
4.1 Element Equations of Motion   77
   4.1.1 Uncondensed Equations   77

    4.1.2 Condensed Equations 78
4.2 Global Equations of Motion 78
4.3 Extensions to Basic Element 79
    4.3.1 General 79
    4.3.2 Consistent Mass Matrix for Open-Type Element 80
        4.3.2.1 Member Mass Matrix 80
        4.3.2.2 Coordinate Transformations 81
        4.3.2.3 Effective Mass Matrix 82
4.4 Modal Superposition Solutions 82
    4.4.1 Free Vibration 82
    4.4.2 Generalized Coordinates 84
    4.4.3 Uniform Base Motion 85
        4.4.3.1 Modification of Equations of Motion 85
        4.4.3.2 Response Spectrum Analysis 87
    4.4.4 Complex Response Method 88
4.5 Direct Integration Solutions 90
4.6 Case Studies 90
    4.6.1 Free Vibration of Cylindrical Shell 90
    4.6.2 Free Vibration of Hemispherical Shell 93
    4.6.3 Dynamic Analysis of Column-Supported Cooling Tower Shell 95
        4.6.3.1 Description of Shell 95
        4.6.3.2 Free Vibration Analysis 96
        4.6.3.3 Response Spectrum Analysis 96
    4.6.4 Cooling Tower on an Interactive Foundation 100
        4.6.4.1 Shell Model 100
        4.6.4.2 Free Vibration Analysis 104
        4.6.4.3 Response Spectrum Analysis 108
        4.6.4.4 Translation and Rocking 112
        4.6.4.5 Interactive Pile Foundation 114
    4.6.5 Cylindrical Shell under Blast Load 122
    4.6.6 Hyperboloidal Shell under Dynamic Wind Load 123
    4.6.7 Free Vibration of Fluid-Filled Cylinder 123

Appendix 127

References 127

5 Geometric Nonlinearity and Instability 130
5.1 General 130
5.2 Strain–Displacement Relations 130
5.3 Modified Equilibrium Equations 131

5.4 Geometric Stiffness Matrix 132
    5.4.1 Definition 132
    5.4.2 Transformation Matrix 133
    5.4.3 Evaluation 133
5.5 Displacement Interpolation Functions 135
    5.5.1 General Considerations 135
    5.5.2 First-order Interpolations 135
    5.5.3 Higher-order Interpolations 138
5.6 Bifurcation Buckling 138
    5.6.1 Procedure 138
    5.6.2 Hyperbolic Cooling Tower Shell 139
    5.6.3 Effect of Imperfections 141
5.7 Incremental Nonlinear Analysis 143

References 145

6 Analysis of Locally Non-Axisymmetric Shells 147
6.1 Concept 147
6.2 General Shell Element 149
6.3 Transitional Element 151
6.4 Global System 152
6.5 Solution Procedure 154
6.6 Case Studies 157
    6.6.1 Imperfect Hyperboloidal Shell under Self-Weight and Wind Loading 157
    6.6.2 Cylindrical Shell with a Circular Cut-Out 164
6.7 Extensions 166

References 168

7 Computer Programs and Case Study 169
7.1 General 169
7.2 Computer Programs 169
    7.2.1 BOSOR4 169
    7.2.2 SHORE III 171
7.3 Case Study of a Hyperboloidal Shell on Column Supports 173
    7.3.1 Scope 173
    7.3.2 Geometry 174
    7.3.3 Discretization Pattern 175
    7.3.4 Loading 177
    7.3.5 Self Weight Stress Analysis 181
    7.3.6 Wind Load Stress Analysis 185

7.3.7 Earthquake Load Stress Analysis 187
7.3.8 Influence of Upper Ring Beam 187

References 199

Index 201

Preface

This monograph focuses on the finite element analysis of thin shells of revolution as a distinct geometrical and structural form. The main thrust is to treat each component in the analysis, e.g., geometry, loading, mass, in the most precise analytical manner. In turn, this approach leads to more efficient computer-based numerical solutions. The emphasis is on linear problems, both static and dynamic, since most engineering applications fall within this regime; however, many of the concepts are applicable to nonlinear problems, as well. Considerable emphasis is placed on the techniques of modeling, with detailed discussion of discretization, representation of loading and interpretation of results. This is carried out mainly through commentaries within the numerous case studies.

The concepts presented have been implemented in a computer program, SHORE, which has been continually expanded and upgraded. This has enabled a variety of example problems to be solved and presented. However, the development is intended to be more general than simply a theoretical manual for a specific program. In fact, it is felt that the focus on a high-precision formulation will be even more relevant for the next generation of microcomputer-based finite element codes.

The book is largely based on investigations carried out at Washington University by the author and several collaborators. Most of the material first appeared in the theses or dissertations of Lawrence J. Brombolich, Subir K. Sen, Herman B. Suryoutomo, Richard D. Lowrey, Osama M. El-Shafee, P. K. Basu, Kye J. Han and Bor-Jen Lee. Additional studies have been performed by Mr Jhun-Sou Lin. Much of the work has been facilitated by research supported by the National Science Foundation under programs monitored by Michael Gaus and John Goldberg.

The author has drawn upon many outside sources as well. The writings of David Bushnell and John Abel have been particularly valuable.

The efforts of Ms Kathy Schallert in typing the manuscript and Mr Long Phan in preparing the illustrations are appreciated.

Phillip L. Gould

1 Introduction

1.1 Description of structural form

A shell is a continuum which is bounded by two curved surfaces separated by the thickness. When the thickness is considerably smaller than the principal radii of curvatures of the bounding surfaces, the shell is said to be *thin*. If each bounding surface is generated by the rotation of a plane curve about a common axis, a *shell of revolution* is produced. The analysis of this structural form is the objective of this study.

1.2 Scope of treatment

Some familiarity with the classical literature is presumed by the author. The reader who desires this background is referred to the author's earlier book in which many of the significant contributions to the vast literature on this subject are noted.[1] Here, the focus is on the numerical stress analysis of rotational shells. However, some justification is required when one turns to a numerical solution, and it seems appropriate to offer a few comments along this line.

Succinctly, the ease or difficulty in obtaining a solution for a shell problem is dependent on the geometry, loading and boundary conditions. *Continuous* geometry, *smooth* loading and *idealized* boundary conditions are generally required to obtain analytical solutions. Deviations from any of the preceding result in complications which may prevent, or greatly complicate, the attainment of an analytical solution, thereby directing the analyst to numerical techniques, the most prominent of which is the *finite element* method.

In this book, a unified methodology is developed for analyzing shells of revolution which may have varying geometry; may be subjected to static, dynamic and/or thermal nonsymmetric loading; and may be supported by discontinuous boundaries. The level of the treatment is intermediate: dynamic and stability formulations are presented and stiffening, layering and branching are considered, in addition to the elementary elastostatic case. On the other hand, more specialized topics such as material non-linearity, creep, and large displacements are not addressed. Somewhat in parallel to this treatment, but with different emphasis, Bushnell[2] has completed a comprehensive review of computerized analysis of rotational shells which covers some of these topics and evaluates the capabilities of several computer programs.

A companion computer program SHORE is available to implement many of the analysis techniques described in the book, but the treatment is fairly independent of the specific program. However, there are choices that need to be made along the way and some of the directions taken in the present development are naturally influenced by experience gained with the SHORE program.

When a numerical approach is pursued, it is productive to re-examine the assumptions of the corresponding analytical problem, with the thought that it may be possible to relax certain of the assumptions and possibly increase the generality of the numerical approach. In this case, there are two such assumptions which bear scrutiny, the use of a reference or *middle surface* to achieve a two-dimensional shell theory and the enforcement of *Kirchhoff's hypothesis* to eliminate transverse shearing strains.

As mentioned in the first sentence of the chapter, a shell is a continuum. As such, it is amenable to treatment by a continuum theory, in this case, the theory of elasticity. There are obvious reasons for defining a middle surface and basing classical shell theory on the displacements of this surface. However, successful finite element models for shells have been developed using the continuum approach, which necessitates coordinates being defined on at least both bounding surfaces rather than only on the middle surface. The choice followed in this book is to generally retain the notion of a middle surface defined *a priori*, but the continuum basis may offer significant advantages for some problems[3] and is used in Chapter 6 for the general transition elements. It should also be noted that for some applications, the choice of an alternate reference surface, perhaps the inside or outside, is preferred.[2]

The approach of classical shell theory is modified somewhat by the inclusion of transverse shearing strains. Kirchhoff's hypothesis, a necessity for obtaining most analytical solutions, is expendable in the finite element method. In fact, it appears that the formulation is improved without the elimination of transverse shearing strains because of a relaxation of the continuity requirements.

In sum, a re-examination of the assumptions of classical shell theory resulted in retaining the middle reference surface while rejecting the elimination of transverse shearing strains.

Introduction

1.3 Historical review

The finite element literature is vast so that, at best, only a very selective review is practical. Concentrating on shells of revolution, the early literature is perhaps a little less voluminous than one might suspect considering the importance and popularity of this structural form. This is probably because, using harmonic analysis, the linear shell equations may be reduced to a one-dimensional problem for which alternative numerical solution methods became feasible with the emergence of high speed digital computers. Amongst the alternatives are numerical integration, finite differences and transfer matrix techniques. Ultimately, in execution, all of these take on the prominent characteristics of finite elements, with stations or intervals being positioned throughout the domain. In fact, the distinction between finite element and finite difference models is practically obliterated when the latter are formulated on an energy basis.[2] Perhaps the most comprehensive rotational shell computer programs, the BOSOR series, which are discussed in Section 7.2.1, are so based.[2,4]

It is perhaps still not universally accepted that the finite element approach is superior for rotational shells and this will not be argued here. However, there is experience to show that the technology transfer from the developer of a computer code to a non-expert user is easiest with finite elements. Furthermore, a finite element formulation may be less sensitive to the selection of various input parameters which affect the stability and accuracy of the analysis. Several commercial and research programs using alternative techniques are very user-dependent for that reason and some require a fair level of expertise to obtain reliable results. Presently, it is thought that the finite element approach is well established for this class of structure and will deliver superior results to the typical informed, but not necessarily expert, user for a wide variety of problems.

The earliest finite elements developed for rotational shells used conical frusta to represent the actual shell.[5-9] However, modeling discontinuities in slope and curvature were introduced into doubly curved shells and it was found that spurious bending moments would be produced in regions where only membrane forces should exist.[10] Improvements were obtained by the development of curved elements[11,12] with the slopes matching those of the actual shell at the nodal circles, while discontinuities of curvature were still disregarded. Ultimately, an element with continuous slopes and curvatures was produced, which essentially eliminated errors due to approximations of the geometry.[13]

It was generally assumed in these studies that the displacements over the entire domain of the shell could be described in terms of displacements at the nodal circles and that the actual loading on the structure could be replaced by circumferential line loads, each of which corres-

3

ponds to a nodal displacement. Non-axisymmetrical loading systems were represented by Fourier series' in the circumferential variable. Correspondingly, the expressions for displacements and stress resultants were taken in similar types of Fourier series', removing the dependence on the polar angle θ. The same basic approach is followed in this book and will be described in detail in the following chapter.

The order of the polynomial approximations for the dependent variables used in early studies was restricted by the number of nodal variables available to evaluate the coefficients. Several possibilities were explored to accommodate higher order polynomials. Pian proposed a refinement whereby additional terms could be added to the comparison functions, with the coefficients determined by minimization of the energy functional.[14] Other authors defined additional nodal variables which produced continuity of an order higher than the minimum required,[15] but the technique originally suggested by Pian was shown to be more efficient.[16]

Another important early contribution was made by Archer in computing *consistent* equivalent nodal forces from distributed loading.[17] Also, several methods for evaluating the stress resultants from the computed displacement fields have been examined.[18] Most of the aforementioned papers have been addressed in an early review[19] and further contributions which impact the present development are cited in the ensuing chapters. A comprehensive bibliography is also contained in Ref. 2.

1.4 Notation for matrices

Matrices are generally designated by brackets [ ], or by bold-face symbols. Diagonal matrices are indicated by special brackets ⌈ ⌋. Column matrices or vectors are represented by { } brackets and row matrices by ⌊ ⌋ brackets, or as the transpose of a column vector, { }^{T}.

Derivatives of matrices with respect to the spatial variable, s, is denoted by [ ]$_{,s}$ and with respect to the time variable, t, by [˙]. Each element of the matrix is to be so differentiated.

References

1. Gould, P. L. *Static Analysis of Shell Structures*, Lexington Press, D. C. Heath and Co., 1977.
2. Bushnell, D., 'Computerized Analysis of Shells-Governing Equations', *J. Computers and Structures*, Vol. 18, No. 3, 1984, pp. 471–536.
3. Ahmad, S., Irons, B. M. and Zienkiewicz, O. C., 'Curved Thick Shell and

Membrane Elements with Particular Reference to Axisymmetric Problems' *Proc. 2nd Conf. on Matrix Methods in Structural Mechanics*, AFFDL-TR-68-150, Wright-Patterson Air Force Base, Dayton, Ohio, 1969, pp. 539–572.
4. Bushnell, D., 'Stress, Stability and Vibration of Complex, Branched Shells of Revolution', *J. Computers and Structures*, Vol. 4, 1974, pp. 399–435.
5. Grafton, P. E. and Strome, D. R., 'Analysis of Axisymmetrical Shells by the Direct Stiffness Method', *AIAA J.*, Vol. 1, No. 10, Oct. 1963, pp. 2342–2347.
6. Hill, D. W. and Coffin, G. K., 'Stresses and Deflections in Cooling Tower Shells Due to Wind Loadings', *Bulletin of I.A.S.S.*, No. 35, Sept. 1968, pp. 45–51.
7. Percy, J. H., Pian, T. H. H., Klein, S. and Navaratna, D. R., 'Application of Matrix Displacement Method to Linear Elastic Analysis of Shells of Revolution', *AIAA J.*, Vol. 3, No. 11, Nov. 1965, pp. 2138–2145.
8. Popov, E. P., Penzien, J. and Lu, Z. A., 'Finite Element Solution for Axisymmetrical Shells', *J. Engrg. Mech. Div., ASCE*, Vol. 90, No. 5, Oct. 1964, pp. 119–145.
9. Zienkiewicz, O. C. and Cheung, Y. K., *The Finite Element Method in Structural and Continuum Mechanics*, McGraw-Hill Book Co. Ltd, London, 1967.
10. Jones, R. E. and Strome, D. R., 'A Survey of Analysis of Shells by the Displacement Method', *Proc. Conf. on Matrix Methods in Structural Mechanics*, AFFDL-TR-66-80, Air Force Flight Dynamics Laboratory, Wright-Patterson AFB, Ohio, 1966, pp. 205–229.
11. Jones, R. E. and Strome, D. R., 'Direct Stiffness Method Analysis of Shells of Revolution Utilizing Curved Elements', *AIAA J.*, Vol. 4, No. 9, Sept. 1966, pp. 1519–1525.
12. Stricklin, J. A., Pian, T. H. H. and Navaratna, D. R., 'Improvements on the Analysis of Shells of Revolution by the Matrix Displacement Method', *AIAA J.*, Vol. 4, No. 11, Nov. 1966, pp. 2069–2071.
13. Khojasteh-Bakht, M., 'Analysis of Elastic–Plastic Shells of Revolution Under Axisymmetric Loading by the Finite Element Method', Structural Engineering Laboratory Report SESM 67-8, University of California, Berkeley, Calif., April 1967.
14. Pian, T. H. H., 'Derivation of Element Stiffness Matrices', *AIAA J.*, Vol. 2, No. 3, Mar. 1964, pp. 576–577.
15. Chan, A. S. L. and Firmin, A., 'The Analysis of Cooling Towers by the Matrix Finite Element Method', *Aeron. J.*, Vol. 74, Oct. 1970, pp. 826–835; Dec., 1970, pp. 971–982.
16. Gould, P. L., Szabo, B. A. and Suryoutomo, H. B., 'Curved Rotational Shell Elements by the Constraint Method', in *Variational Methods in Engineering*, Vol. II, Proc. of an International Conf., University of Southampton, Sept. 1972, pp. 8133–8142.
17. Archer, J. S., 'Consistent Matrix Formulations for Structural Analysis Using Finite-Element Techniques', *AIAA J.*, Vol. 3, No. 10, Oct. 1965, pp. 1910–1918.

Finite element analysis of shells of revolution

18. Navaratna, D. R., 'Computation of Stress Resultants in Finite Element Analysis', Tech. Note, *AIAA J.*, Vol. 4, No. 11, Nov. 1966, pp. 2058–2060.
19. Brombolich, L. J. and Gould, P. L., 'Finite Element Analysis of Shells of Revolution by Minimization of the Potential Energy Functional', Proceedings, Conference on 'Applications of the Finite Element Method in Civil Engineering', Vanderbilt University, Nashville, Tennessee, 13–14 November 1969.

2 Fundamentals

2.1 Surface geometry

Since the surface is axisymmetric, it is convenient to first consider the meridian as a planar curve described in the $R-Z$ Cartesian coordinate system by an equation of the form

$$R = R(Z) \tag{2.1}$$

The pointwise description of the surface is completed by the circumferential coordinate θ, measured in the plane $Z = $ constant, as shown in Fig. 2.1. Thus, any point on the middle surface is uniquely located by the ordered triple (R, θ, Z). In general, the origin $Z = 0$ may be arbitrarily set at a convenient location along the meridian. However, because of the special treatment required for singularities at the pole of a closed shell where $R = 0$, the origin is most conveniently located at the pole for that case.

The principal radii of curvature of the surface, R_ϕ and R_θ, are also shown on Fig. 2.1. They are related by the Gauss–Codozzi equation

$$R_{,\phi} = (R_\theta \sin \phi)_{,\phi} = R_\phi \cos \phi \tag{2.2}$$

It is often expedient to specify the meridian in terms of the angle ϕ as shown in Fig. 2.1 and then to transform between ϕ and Z using

$$R_\phi(Z) = \frac{[1+(R_{,Z})^2]^{3/2}}{R_{,ZZ}} \tag{2.3}$$

$$\sin \phi(Z) = [1+(R_{,Z})^2]^{-1/2} \tag{2.4}$$

and

$$\cos \phi(Z) = R_{,Z}[1+(R_{,Z})^2]^{-1/2} \tag{2.5}$$

where the derivatives of R are determined from Eq. 2.1. Equation 2.3 is familiar from basic surface geometry while Eqs 2.4 and 2.5 are derived in Ref. 1.

Finite element analysis of shells of revolution

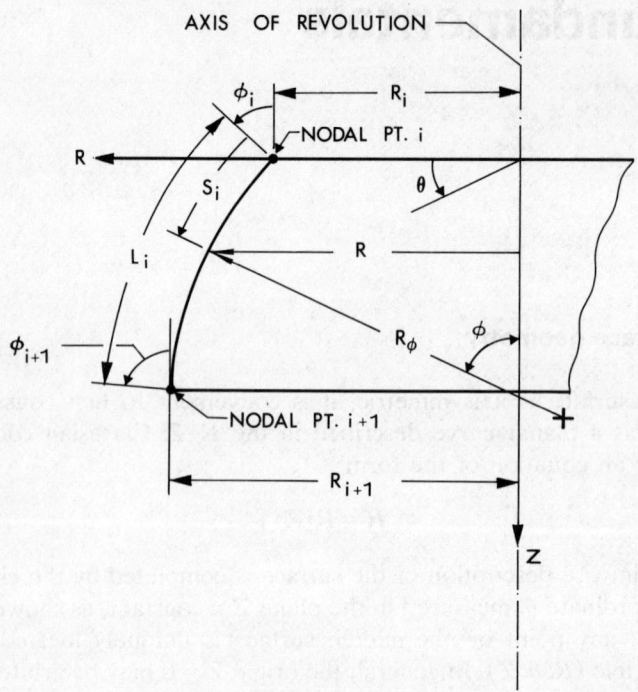

Fig. 2.1 Geometry of the rotational shell element

2.2 Ring element

The shell is discretized into ring elements bounded by two parallel planes which are perpendicular to the axis of rotation. The intersections of these planes with the shell middle surface are called *nodal circles,* as shown in Fig. 2.1, and serve the same purpose as nodal points for two- and three-dimensional finite elements. The length of the meridian of a typical element i between nodal circles i and $i+1$ is

$$L_i = \int_{Z_i}^{Z_{i+1}} [1+(R_{,Z})^2]^{1/2} \, dZ \tag{2.6a}$$

For a closed shell which is quite flat at the pole, the length of the meridian is more accurately computed from

$$L_i = \int_{R_i}^{R_{i+1}} [1+(Z_{,R})^2]^{1/2} \, dR \tag{2.6b}$$

In addition to the Z and ϕ meridional coordinates previously introduced, it is useful to employ a non-dimensional arc length

$$s_i = S_i/L_i \quad (0 \le s_i \le 1) \tag{2.7}$$

such that $s_i(Z_i) = 0$ and $s_i(Z_{i+1}) = 1$. In this case the differentiation formulae

$$(\ )_{,S} = 1/L_i(\ )_{,s} = 1/R_\phi(\ )_{,\phi} \tag{2.8}$$

are helpful. The subscript will be dropped from s_i unless it is explicitly required.

In the course of the derivation of the element matrices, numerous numerical integrations are required due to the presence of a number of geometric parameters which are conveniently identified by P_k ($k = 1, 2, \ldots, 14$):

$$P_1 = R \qquad P_2 = \sin\phi \qquad P_3 = \cos\phi \qquad P_4 = \frac{1}{R}$$

$$P_5 = \frac{\sin\phi}{R} \qquad P_6 = \frac{\cos\phi}{R} \qquad P_7 = \frac{\sin^2\phi}{R} \qquad P_8 = \frac{\cos^2\phi}{R}$$

$$P_9 = \frac{\sin\phi\cos\phi}{R} \qquad P_{10} = \frac{1}{R_\phi} \qquad P_{11} = \frac{R}{R_\phi} \qquad P_{12} = \frac{\sin\phi}{R_\phi}$$

$$P_{13} = \frac{\cos\phi}{R_\phi} \qquad P_{14} = \frac{R}{R_\phi^2} \tag{2.9}$$

It has been found to be convenient to express each of these parameters $P_k(s)$ ($k = 1, \ldots, 14$) in the form of a fourth-order Lagrangian polynomial

$$P_k(s) = \bar{P}_{k0} + \bar{P}_{k1}s + \bar{P}_{k2}s^2 + \bar{P}_{k3}s^3 + \bar{P}_{k4}s^4 \tag{2.10}$$

which matches at the end points, $P_k(0)$ and $P_k(1)$, the mid-point, $P_k(0.5)$, and the quarter-points, $P_k(0.25)$ and $P_k(0.75)$, of the meridian. The weighting functions P_{kj} ($j = 0, \ldots, 4$) are given by[2]

$$\begin{Bmatrix} \bar{P}_{k0} \\ \bar{P}_{k1} \\ \bar{P}_{k2} \\ \bar{P}_{k3} \\ \bar{P}_{k4} \end{Bmatrix} = \begin{bmatrix} 1 & 0 & 0 & 0 & 0 \\ -25/3 & 16 & -12 & 16/3 & -1 \\ 70/3 & -208/3 & 76 & -112/3 & 22/3 \\ -80/3 & 96 & -128 & 224/3 & -16 \\ 32/3 & -128/3 & 64 & -128/3 & 32/3 \end{bmatrix} \begin{Bmatrix} P_k(0) \\ P_k(0.25) \\ P_k(0.50) \\ P_k(0.75) \\ P_k(1.0) \end{Bmatrix}$$

(2.11)

In the case of a closed shell, several of the geometric parameters such as $1/R$ and $\cos\phi/R$ apparently become singular at $s = 0$. However, in the resulting integrals, such functions are always combined in the integrands with another term such as s or s^2 which also tends to zero. The resulting indeterminate form may be evaluated using L'Hospital's rule and then substituted for $P_k(0)$ in Eq. 2.10.

Since the total number of definitive parameters may become quite large, particularly for variable thickness shells, a further list is not given here. The geometrical representation would generally be handled automatically by the computer code.

Occasionally, rotational shells are encountered which are defined only by a plotted curve $R = R(Z)$, or by tabulated points along the meridian without a mathematical equation in the form of Eq. 2.1. Such cases occur in imperfect shells where significant deviations may be measured, or in rotationally symmetric objects which have a complicated, non-analytically defined shape due to functional or natural reasons.

The main geometrical problem encountered in such cases is in the evaluation of the $P_k(s)$ terms when derivatives such as $R_{,Z}$ and $R_{,ZZ}$ are involved. If a high-order curve is fitted to a number of scaled points, the computed derivatives are likely to vary rapidly and invalidate the analysis. Hermitian polynomials, in which derivatives as well are matched, are limited by the difficulty of precisely determining the derivatives of the plotted or tabulated curve pointwise. A more reliable approach is to use a relatively low-order, perhaps quadratic or cubic, least squares fit and to then derive the derivatives accordingly from the analytical expression.

Although several other alternatives for representing the geometry are possible, the approach presented in this section permits subsequent integrals to be evaluated in closed form. Moreover, only exact geometric data is introduced and differentiation of interpolation polynomials, a possible source of inaccuracies, is avoided.

2.3 Kinematic and statical variables

2.3.1 Definitions

Numerous dependent variables enter into the shell of revolution problem and it is advantageous to carefully define and delineate between the various sets. Following the custom of shell theory, these variables are referred to the curvilinear coordinates ϕ and θ. It is convenient to include the possibility of variation with time, t, as well as space.

(1) Middle surface displacements, D

$$\{D(s, \theta; t)\} = \{u \, v \, w \, \beta_\phi \, \beta_\theta\} \tag{2.12}$$

in which u, v and w are the meridional, circumferential and normal displacements; and β_ϕ and β_θ are the meridional and circumferential rotations, positive as shown in Fig. 2.2.

Fundamentals

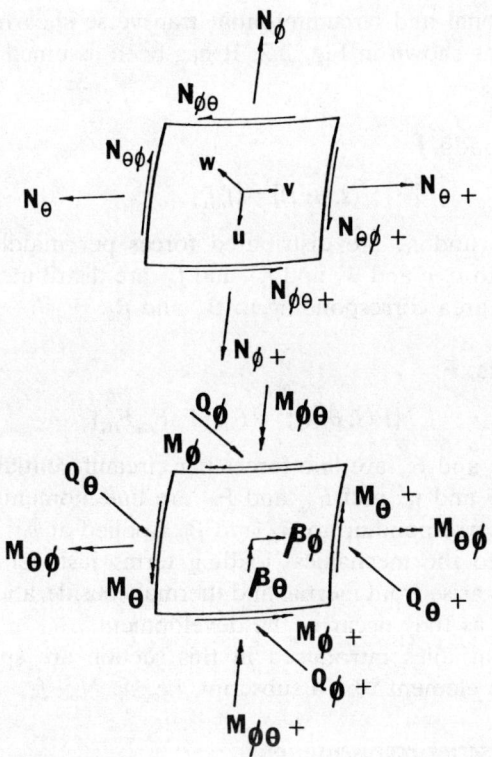

Fig. 2.2 Sign convention

(2) Middle surface strains and changes in curvature, ε

$$\{\varepsilon(s,\theta;t)\} = \{\varepsilon_\phi \varepsilon_\theta \varepsilon_{\theta\phi} \kappa_\phi \kappa_\theta \kappa_{\theta\phi} \gamma_\phi \gamma_\theta\} \quad (2.13)$$

in which ε_ϕ, ε_θ, and $\varepsilon_{\theta\phi}$ are the meridional and circumferential extensional strains and the in-plane shearing strain; κ_ϕ, κ_θ and $\kappa_{\theta\phi}$ are the meridional and circumferential changes in curvature, and the twist; and γ_ϕ and γ_θ are the meridional and circumferential transverse shearing strains.[1]

(3) Stress resultants and stress couples, **N**

$$\{N(s,\theta;t)\} = \{N_\phi N_\theta N_{\theta\phi} M_\phi M_\theta M_{\theta\phi} Q_\phi Q_\theta\} \quad (2.14)$$

in which N_ϕ, N_θ, and $N_{\theta\phi}$ are the meridional and circumferential extensional stress resultants and the in-plane shearing stress resultant (sometimes called membrane stress resultants); M_ϕ, M_θ, and $M_{\theta\phi}$ are the meridional, circumferential and twisting stress couples; and Q_ϕ and Q_θ

are the meridional and circumferential transverse shearing stress resultants, positive as shown in Fig. 2.2. It has been assumed that $N_{\phi\theta} = N_{\theta\phi}$ and $M_{\phi\theta} = M_{\theta\phi}$.

(4) Surface loads, f

$$\{f(s, \theta; t)\} = \{f_u f_v f_w f_{\beta_\phi} f_{\beta_\theta}\} \tag{2.15}$$

in which f_u, f_v and f_w are distributed forces per middle surface area corresponding to u, v and w; and f_{β_ϕ} and f_{β_θ} are distributed moments per middle surface area corresponding to β_ϕ and β_θ.

(5) Ring loads, F

$$\{F(\bar{s}, \theta; t)\} = \{F_u F_v F_w F_{\beta_\phi} F_{\beta_\theta}\} \tag{2.16}$$

in which F_u, F_v and F_w are line forces per circumferential length corresponding to u, v and w; and F_{β_ϕ} and F_{β_θ} are line moments per circumferential length corresponding to β_ϕ and β_θ, applied at $s = \bar{s}$ ($\bar{s} = 0$ or 1).

In addition to the mechanical loading terms just defined, additional load type terms arise from inertial and thermal considerations. Such terms will be defined as they occur in the development.

All of the variables introduced in this section are specialized for a particular finite element i by a subscript, i.e. u_i, $N_{\phi i}$, f_{wi}.

2.3.2 *Fourier series representation*

The analysis of rotationally symmetric bodies, in general, and shells of revolution, in particular, is expedited by the expansion of all dependent variables into Fourier series' in the circumferential variable θ. This achieves *separation* of the independent variables and the two-dimensional problem is reduced to a series of one-dimensional problems. Moreover, in a linear formulation, the governing equations are uncoupled and each harmonic may be solved independently. In most cases, a partial Fourier series for each variable ($\sin j\theta$ or $\cos j\theta$) is adequate; in the rare exception, the partial series is easily generalized. Also, time dependence may be included when required.

(1) Middle surface displacements, $\mathbf{\Theta}_1^j \mathbf{D}^j$

$$\{D\} = \sum_{j=0}^{\infty} \lceil \Theta_1^j(\theta) \rfloor \{D^j(s)\} \tag{2.17}$$

in which

$$\lceil \Theta_1^j \rfloor = \lceil \cos j\theta \; \sin j\theta \; \cos j\theta \; \cos j\theta \; \sin j\theta \rfloor \tag{2.18}$$

and
$$\{D^j\} = \{u^j v^j w^j \beta_\phi^j \beta_\theta^j\} \quad (2.19)$$

(2) Middle surface strains and changes in curvature, $\Theta_2^j \varepsilon^j$

$$\{\varepsilon\} = \sum_{j=0}^{\infty} \{\varepsilon\}^j = \sum_{j=0}^{\infty} \lceil \Theta_2^j(\theta) \rfloor \{\varepsilon^j(s)\} \quad (2.20)$$

in which

$$\lceil \Theta_2^j \rfloor = \lceil \cos j\theta \cos j\theta \sin j\theta \cos j\theta \cos j\theta \sin j\theta \cos j\theta \sin j\theta \rfloor \quad (2.21)$$

and

$$\{\varepsilon^j\} = \{\varepsilon_\phi^j \varepsilon_\theta^j \varepsilon_{\theta\phi}^j \kappa_\phi^j \kappa_\theta^j \kappa_{\theta\phi}^j \gamma_\phi^j \gamma_\theta^j\} \quad (2.22)$$

(3) Stress resultants and stress couples, $\Theta_2^j N^j$

$$\{N\} = \sum_{j=0}^{\infty} \lceil \Theta_2^j(\theta) \rfloor \{N^j(s)\} \quad (2.23)$$

in which

$$\{N^j\} = \{N_\phi^j N_\theta^j N_{\theta\phi}^j M_\phi^j M_\theta^j M_{\theta\phi}^j Q_\phi^j Q_\theta^j\} \quad (2.24)$$

(4) Surface loads, $\Theta_1^j f^j$

$$\{f\} = \sum_{j=0}^{\infty} \lceil \Theta_1^j(\theta) \rfloor \{f^j(s)\} \quad (2.25)$$

in which

$$\{f^j\} = \{f_u^j f_v^j f_w^j f_{\beta_\phi}^j f_{\beta_\theta}^j\} \quad (2.26)$$

(5) Ring loads, $\Theta_1^j f^j$

$$\{F\} = \sum_{j=0}^{\infty} \lceil \Theta_1^j(\theta) \rfloor \{F^j\} \quad (2.27)$$

in which

$$\{F^j\} = \{F_u^j F_v^j F_w^j F_{\beta_\phi}^j F_{\beta_\theta}^j\} \quad (2.28)$$

Applied loading in the form of concentrated forces and moments may be represented *exactly* in the *meridional* direction by these ring loads, and *approximately* in the *circumferential* direction by a uniform or sinusoidal distribution over a short arc. This distributed loading is then expanded in a Fourier series in the form of Eq. 2.27. An example for a column-supported shell will be given later.

Finite element analysis of shells of revolution

It should be noted that in most cases, the number of Fourier components will be fewer for the $j = 0$ harmonic. Except for rare circumstances such as global torsion of the shell, v^0 and β_θ^0 along with the corresponding force terms will be zero and the resulting matrices will be reduced accordingly. For the most part, $j \geq 1$ will be treated as the general case.

2.4 Kinematic laws

2.4.1 Transverse shearing strains

In the classical analysis of shells, the set of displacements, $\{D\}$ is almost always contracted from five to four terms by the suppression of the transverse shearing strains γ_ϕ and γ_θ, which also eliminates β_θ. This is known as Kirchhoff's hypothesis and produces significant mathematical simplifications in most cases. However, in the finite element approach, the corresponding advantages are not apparent. Although a reduction in the total number of displacements is again achieved by applying Kirchhoff's hypothesis, a differentiated term ($w_{,s}$) is introduced into the set and this is disadvantageous in a numerical formulation, as will be elaborated in Section 3.1.1. Therefore, Kirchhoff's hypothesis is *not* invoked in this development. In other words, *transverse shearing strains are retained* in the formulation, not because they are necessarily significant but primarily to achieve a more desirable interelement continuity requirement, which is also explained in more detail in Section 3.1.1. It should be noted that regardless of the enforcement of Kirchhoff's hypothesis, the transverse shearing forces must be retained. In the absence of corresponding strains, they are computed from the moment gradients, analogous to elementary beam theory (see Eqs. 2.71 and 2.72).

It is of interest to note that recent studies using classical shell equations revealed that the proper inclusion of transverse shearing strain enables solutions for shells subject to concentrated loads (which obviously mobilize transverse shear effects) to be obtained.[3] It is therefore expected that the element under consideration will accommodate such intense loadings better than those based on Kirchhoff's hypothesis. Further, for shells formed from laminated or composite materials, the transverse shearing strains may be significant even for distributed loadings.[4]

2.4.2 Linear form

A general form of the kinematic law, relating the strains and displacements, is

$$\{\varepsilon\} = [A(s)]\{D\} \qquad (2.29)$$

Fundamentals

In the case of linear shell theory, the operator matrix $[A(s)]$ does not depend on $\{D\}$. For a general rotational element $(R \neq 0)$ and harmonic j, we have

$$\{\varepsilon^j\} = [A^j(s)]\{D^j\} \tag{2.30}$$

in which[1,2]

$$[A^j(s)] = \begin{bmatrix} (1/L_i)(\ )_{,s} & 0 & 1/R_\phi & 0 & 0 \\ \cos\phi/R & j/R & \sin\phi/R & 0 & 0 \\ -j/R & \begin{bmatrix}(1/L_i)(\ )_{,s} \\ -\cos\phi/R\end{bmatrix} & 0 & 0 & 0 \\ 0 & 0 & 0 & (1/L_i)(\ )_{,s} & 0 \\ 0 & 0 & 0 & \cos\phi/R & j/R \\ 0 & 0 & 0 & -j/2R & \begin{bmatrix}(1/2L_i)(\ )_{,s} \\ -\cos\phi/2R\end{bmatrix} \\ -1/R_\phi & 0 & (1/L_i)(\ )_{,s} & 1 & 0 \\ 0 & -\sin\phi/R & -j/R & 0 & 1 \end{bmatrix} \tag{2.31}$$

and $(\ )_{,s}$ stands for the appropriate element of $\{D^j\}$ which is to be differentiated.

Next, we consider the closed element, which may be a dome $[\phi(R = 0) = 0]$, to be numbered $i = 0$. For those expressions which became singular at $r = 0$ (rows 2, 3, 5, 6 and 8 of the preceding matrix) alternate kinematic relationships are derived by evaluating the limits as $r \to 0$ using L'Hospital's rule. A number of relationships between the pole displacements may be derived and an alternate equation written:

$$\{\varepsilon^j\} = [A^j(0)]\{D^j(0)\} \tag{2.32}$$

in which

$$[A^j(0)] = \begin{bmatrix} (1/L_0)(\ )_{,s} & 0 & 1/R_\phi(0) & 0 & 0 \\ (1/L_0)(\ )_{,s} & (j/L_0)(\ )_{,s} & 1/R_\phi(0) & 0 & 0 \\ (-j/L_0)(\ )_{,s} & 0 & 0 & 0 & 0 \\ 0 & 0 & 0 & (1/L_0)(\ )_{,s} & 0 \\ 0 & 0 & 0 & (1/L_0)(\ )_{,s} & (j/L_0)(\ )_{,s} \\ 0 & 0 & 0 & (-j/2L_0)(\ )_{,s} & 0 \\ -1/R_\phi(0) & 0 & (1/L_0)(\ )_{,s} & 1 & 0 \\ 0 & -1/R_\phi(0) & (-j/L_0)(\ )_{,s} & 0 & 1 \end{bmatrix} \tag{2.33}$$

Finite element analysis of shells of revolution

In Eq. 2.33 the differentiated terms are to be evaluated at $R = 0$. For a dome, L_0 should be computed using Eq. 2.6(b) with $R_i = 0$ and $R_{i+1} = R_1$.

2.4.3 Nonlinear form

Finally, we are interested in the case where moderate rotations may occur along with the small deformations covered by the preceding expressions. The corresponding strain–displacement relationships are a necessary component of a linear stability analysis. We generalize Eq. 2.29 into[5]

$$\{\hat{\varepsilon}\} = \{\varepsilon\} + \{\check{\varepsilon}\} \tag{2.34}$$

in which

$$\{\check{\varepsilon}\} = \sum_{j=0}^{\infty} \{\check{\varepsilon}\}^j = \sum_{j=0}^{\infty} \tfrac{1}{2}[\check{A}]^j \{\check{D}\}^j \tag{2.35a}$$

and

$$\{\check{\varepsilon}\} = \{\check{\varepsilon}_\phi \check{\varepsilon}_\theta \check{\varepsilon}_{\theta\phi} \check{\kappa}_\phi \check{\kappa}_\theta \check{\kappa}_{\theta\phi} \check{\gamma}_\phi \check{\gamma}_\theta\} \tag{2.35b}$$

The additional strain–displacement terms may be separated harmonically as

$$[\check{A}]^j = [\Theta_3^j][\check{A}^j(s)] \tag{2.36}$$

in which

$$[\Theta_3^j] = \begin{bmatrix} \bar{\Theta}_3^j & 0 \\ 0 & \bar{\Theta}_3^j \\ 0 & 0 \end{bmatrix}_{8\times 8} \tag{2.37}$$

$$[\bar{\Theta}_3^j] = \begin{bmatrix} \cos j\theta & 0 & 0 & 0 \\ 0 & \sin j\theta & 0 & 0 \\ 0 & 0 & \sin j\theta & \cos j\theta \end{bmatrix}_{3\times 4} \tag{2.38}$$

and

$$[\check{A}^j(s)]^T = \begin{bmatrix} \check{\beta}_\phi^j & 0 & \check{\beta}_\theta^j & 0 & 0 & 0 & 0 & 0 \\ 0 & \check{\beta}_\theta^j & \check{\beta}_\phi^j & 0 & 0 & 0 & 0 & 0 \\ 0 & 0 & 0 & \dfrac{2}{R_\phi}\check{\beta}_\phi^j & 0 & \dfrac{2}{R_\phi}\check{\beta}_\theta^j & 0 & 0 \\ 0 & 0 & 0 & 0 & \dfrac{2}{R_\theta}\check{\beta}_\theta^j & \dfrac{2}{R_\theta}\check{\beta}_\phi^j & 0 & 0 \end{bmatrix}_{4\times 8} \tag{2.39}$$

where

$$\check{\beta}_\phi^j = \beta_\phi^j - \gamma_\phi^j \tag{2.40}$$

and
$$\check{\beta}_\theta^j = \beta_\theta^j - \gamma_\theta^j \quad (2.41)$$

Also,
$$\{\check{D}\}^j = \lceil \Theta_4 \rfloor \{\bar{D}^j(s)\} \quad (2.42)$$
$$\lceil \Theta_4^j \rfloor = \lceil \cos j\theta \sin j\theta \cos j\theta \sin j\theta \rfloor \quad (2.43)$$

and
$$\{\bar{D}^j(s)\} = \{\check{\beta}_\phi^j \check{\beta}_\theta^j \beta_\phi^j \beta_\theta^j\}. \quad (2.44)$$

Note that there are no nonlinear terms associated with the transverse shearing strains, but that elements of $\{D^j\}$ are contained in $[\check{A}^j]$, in contrast to $[A^j]$.

Bushnell[6] has observed that the nonlinear terms associated with the changes in curvature are not as important as those associated with the strains, because the *linear* terms in the strain–displacement equations tend to cancel each other for large displacements, while those in the change in curvature–displacement equations do not cancel. As a result, he recommends dropping the nonlinear terms in the curvature expressions. This affects the prebuckled stress matrix as discussed in Section 5.4.3.

2.5 Constitutive laws

One of the great benefits of the finite element approach is the possibility of accommodating shells with complicated material properties. The general form of the relationship between the stress resultants and couples, and the strains and changes in curvature is

$$\{N\} = [H]\{\varepsilon\} - \{N_T\} \quad (2.45)$$

or, for harmonic j,

$$\{N\}^j = \lceil \theta_2^j \rfloor \{N^j\} = \lceil \theta_2^j \rfloor [[H]\{\varepsilon^j\} - \{N_T^j\}] \quad (2.46)$$

in which $[H]$ is an 8×8 matrix dependent on the specific material law and $\{N_T^j\}$ is a thermal stress vector in the form

$$\{N_T^j\} = \{N_{T\phi}^j N_{T\theta}^j\ 0\ M_{T\phi}^j M_{T\theta}^j\ 0\ 0\ 0\}. \quad (2.47)$$

In Table 2.1 the elements of the symmetrical matrix $[H]$ are given for three cases, isotropic and orthotropic single-layered shells and symmetrical multilayered (orthotropic) shells,[7] but the procedure can also treat other cases using appropriate stress–strain laws and evaluating the corresponding elements of $[H]$.

In Table 2.2 the elements of the thermal stress vector are given for the corresponding three cases.

Finite element analysis of shells of revolution

Table 2.1

[H]	Single-layered shells		Symmetrical multilayered shells
	Isotropic	*Orthotropic*	
H_{11}	Hh	$H_0 h$	$\sum_k (H_0)_k h_k$
H_{12}	μHh	$\mu_{\phi\theta} H_0 h$	$\sum_k (\mu_{\phi\theta} H_0)_k h_k$
H_{22}	Hh	$(\mu_{\phi\theta}/\mu_{\theta\phi}) H_0 h$	$\sum_k [(\mu_{\phi\theta}/\mu_{\theta\phi}) H_0]_k h_k$
H_{33}	Gh	$G_{\phi\theta} h$	$\sum_k (G_{\phi\theta})_k h_k$
H_{44}	$Hh^3/12$	$H_0 h^3/12$	$\sum_h (H_0)_k (\zeta_k^2 + h_k^2/12) h_k$
H_{45}	$\mu Hh^3/12$	$\mu_{\phi\theta} H_0 h^3/12$	$\sum_h (\mu_{\phi\theta} H_0)_k (\zeta_k^2 + h_k^2/12) h_k$
H_{55}	$Hh^3/12$	$(\mu_{\phi\theta}/\mu_{\theta\phi}) H_0 h^3/12$	$\sum_h [(\mu_{\phi\theta}/\mu_{\theta\phi}) H_0]_k (\zeta_k^2 + h_k^2/12) h_k$
H_{66}	$Gh^3/6$	$G_{\phi\theta} h^3/6$	$2\sum_h (G_{\phi\theta})_k (\zeta_k^2 + h_k^2/12) h_k$
H_{77}	λGh	$\lambda G_{\phi n} h$	$\lambda \sum_h (G_{\phi n})_k h_k$
H_{88}	λGh	$\lambda G_{\theta n} h$	$\lambda \sum_h (G_{\theta n})_k h_k$

h = thickness
$H = E/(1-\mu^2)$
E, E_ϕ = Young's moduli
$\mu, \mu_{\phi\theta}, \mu_{\theta\phi}$ = Poisson's ratios
λ = shear factor
$H_0 = E_\phi/(1 - \mu_{\theta\phi}\mu_{\phi\theta})$
$G, G_{\phi n}, G_{\theta n}, G_{\phi\theta}$ = shear moduli
h_k = thickness of layer k
$(H_0)_k = H_0$ for layer k
ζ_k = distance from centroid of layer k to middle surface

In addition to shells with varying material properties discussed in this section, this finite element model can conveniently treat certain types of structures which are not shells by strict definition. Some examples are shells with stiffeners, undulations or folds added to supplement the strength and/or rigidity of the basic constant thickness section; and *reticulated* shells which are primarily composed of intersecting grids of closely-spaced members following the topography of a curved surface. By carrying out the formulation at the level of the stress *resultant*–strain relationships, Eq. 2.45, rather than at the stress–strain level used in the Theory of Elasticity, useful forms of [H] have been derived which permit such structures to be analyzed as rotational shells.[1] In particular, it is possible to consider separate effective thicknesses for the extensional and

Table 2.2

$\{N_T^i\}$	Single-layered shells		Symmetrical multilayered shells
	Isotropic	Orthotropic	
$N_{T\phi}^i$	$H(1+\mu)\bar{\alpha}T_1$	$H_0[\bar{\alpha}_\phi + \mu_{\phi\theta}\bar{\alpha}_\theta]T_1$	$\sum_k (H_0)_k[\bar{\alpha}_\phi + \mu_{\phi\theta}\bar{\alpha}_\theta]_k T_{1k}$
$N_{T\theta}^i$	$H(1+\mu)\bar{\alpha}T_1$	$H_0[\mu_{\phi\theta}\bar{\alpha}_\phi + (\mu_{\phi\theta}/\mu_{\theta\phi})\bar{\alpha}_\theta]T_1$	$\sum_k (H_0)_k[\mu_{\phi\theta}\bar{\alpha}_\phi + (\mu_{\phi\theta}/\mu_{\theta\phi})\bar{\alpha}_\theta]_k T_{1k}$
$M_{T\phi}^i$	$H(1+\mu)\bar{\alpha}T_2$	$H_0[\bar{\alpha}_\phi + \mu_{\phi\theta}\bar{\alpha}_\theta]T_2$	$\sum_k (H_0)_k[\bar{\alpha}_\phi + \mu_{\phi\theta}\bar{\alpha}_\theta]_k T_{2k}$
$M_{T\theta}^i$	$H(1+\mu)\bar{\alpha}T_2$	$H_0[\mu_{\phi\theta}\bar{\alpha}_\phi + (\mu_{\phi\theta}/\mu_{\theta\phi})\bar{\alpha}_\theta]T_2$	$\sum_k (H_0)_k[\mu_{\phi\theta}\bar{\alpha}_\phi + (\mu_{\phi\theta}/\mu_{\theta\phi})\bar{\alpha}_\theta]_k T_{2k}$

$\bar{\alpha}, \bar{\alpha}_\phi \bar{\alpha}_\theta$ = coefficients of thermal expansion

T^i = temperature difference Fourier coefficient

$$T_1 = \int_{-h/2}^{h/2} T^i(\zeta)\, d\zeta \qquad T_{1k} = \int_{-h_k/2}^{h_k/2} T^i(\zeta)\, d\zeta$$

$$T_2 = \int_{-h/2}^{h/2} T^i(\zeta)\zeta\, d\zeta \qquad T_{2k} = \int_{-h_k/2}^{h_k/2} T^i(\zeta)\zeta_k\, d\zeta$$

bending terms in orthogonal directions by direct insertion of terms into the $[H]$ matrix. This is illustrated in Section 3.3.2.

2.6 Boundary conditions

A finite element formulation conveniently provides for the enforcement of boundary conditions on those variables which are explicitly contained in the governing equations. The procedure is to simply eliminate (or 'zero') the particular (displacement) variable at the nodal circle where it is constrained. This is usually performed at the global level but is conveniently discussed here. Natural boundary conditions on (force) variables which are not explicit in the governing equations are applied through ring loads $\{F\}$, as defined in Eq. 2.16. It is also possible to specify mixed boundary conditions, where the forces are known on part of the boundary and the displacements are prescribed on the remainder. For example, this is required for shells supported on circumferentially spaced columns.

Specified conditions may be written by evaluating Eqs 2.12 and 2.16 at the boundary, $s = \bar{s}$ ($\bar{s} = 0$ or 1):

$$\{D(\bar{s}, \theta; t)\} = \{\bar{u}\bar{v}\bar{w}\bar{\beta}_\phi\bar{\beta}_\theta\} \tag{2.48}$$

$$\{F(\bar{s}, \theta; t)\} = \{\bar{F}_u\bar{F}_v\bar{F}_w\bar{F}_{\beta_\phi}\bar{F}_{\beta_\theta}\} \tag{2.49}$$

and are given in Table 2.3.

Table 2.3

Condition at $s = \bar{s}$	Specified quantities	Symbol
Free	$\bar{F}_u = \bar{F}_v = \bar{F}_w = \bar{F}_{\beta_\phi} = \bar{F}_{\beta_\theta} = 0$	
Fixed	$\bar{u} = \bar{v} = \bar{w} = \bar{\beta}_\phi = \bar{\beta}_\theta = 0$	
Hinged	$\bar{F}_{\beta_\phi} = 0;\ \bar{u} = \bar{v} = \bar{w} = \bar{\beta}_\theta = 0$	
Roller	$\bar{F}_w = \bar{F}_{\beta_\phi} = 0;\ \bar{u} = \bar{v} = \bar{\beta}_\theta = 0$	
Sliding	$\bar{F}_u = 0;\ u = v = \bar{\beta}_\phi \bar{\beta}_\theta = 0$	

For closed shells, the use of a special cap element supplants the need for boundary conditions at the pole.

Finally, rigid body displacements must be eliminated. This is only relevant for harmonics $j = 0$ and $j = 1$ since the harmonics $j > 1$ are self-equilibrated. A sufficient condition to eliminate rigid body displacements is $\bar{u} = \bar{v} = 0$ on *one edge* of the shell. This is also necessary for shells of negative Gaussian curvature but may be relaxed slightly for shells of positive and zero Gaussian curvature.[8]

2.7 Variational principles

2.7.1 *Hamilton's principle*

Hamilton's Variational Principle (HVP) is sufficiently general to derive the field equations for the static and dynamic response of shells and provides a convenient format for introducing the approximations associated with the finite element displacement method.

The principle is based on the functional[9]

$$L = \int_{t_0}^{t_1} (\Pi_p - K_E)\, dt = \int_{t_0}^{t_1} \bar{L}\, dt \qquad (2.50)$$

in which

$$\Pi_p = \int_{\mathscr{S}} \{N\}^T\{\varepsilon\}\, d\mathscr{S} - \int_{\mathscr{S}} \{f\}^T\{D\}\, d\mathscr{S} - \sum_{\mathscr{S}} \int_\sigma \{F(\sigma)\}\{D(\sigma)\}\, d\sigma \qquad (2.51)$$

= total potential energy of the system

$$K_E = \frac{1}{2} \int_{\mathscr{V}} \rho \{\dot{q}\}^T\{\dot{q}\}\, d\mathscr{V} \qquad (2.52)$$

= kinetic energy of the system

where
$$\rho = \text{mass density}$$
and

$\{q\}$ = displacement vector at *any point* $\left(-\dfrac{h}{2} \leq \zeta \leq \dfrac{h}{2}\right)$ within or on the surface of the shell; $\{q\}$ may be related to the corresponding displacements at the reference surface $\{D\}$ as shown in Section 4.1.1.

Also, $\mathscr{S}$ and $\mathscr{V}$ = the area and volume of the reference surface and σ indicates a circumferential ring or boundary with line loading $\{F(\sigma)\}$ and displacement $\{D(\sigma)\}$.

The stationary condition is given by

$$\delta L = 0 \tag{2.53}$$

It is usually preferable to perform the variation prior to the time integration since these operations are normally interchangeable. Therefore, the stationary condition becomes

$$\int_{t_0}^{t_1} \delta \bar{L}\, dt = 0 \tag{2.54}$$

Proceeding with $\delta \bar{L}$,

$$\delta \Pi_p = \int_{\mathscr{S}} \{N\}^T \delta(\varepsilon)\, d\mathscr{S} - \int_{\mathscr{S}} \{f\}^T \delta\{D\}\, d\mathscr{S} - \int_{\sigma} \{F(\sigma)\}^T \delta\{D(\sigma)\}\, d\sigma \tag{2.55}$$

Now, it is convenient to introduce the constitutive law in the form of Eq. 2.45 and the kinematic law in the form of Eq. 2.29 into Eq. 2.55, so that

$$\delta \Pi_p = \int_{\mathscr{S}} \{AD\}^T [H]\, \delta\{AD\}\, d\mathscr{S} - \int_{\mathscr{S}} \{N_T\}^T\, \delta\{AD\}\, d\mathscr{S}$$
$$- \int_{\mathscr{S}} \{f\}^T\, \delta\{D\}\, d\mathscr{S} - \int_{\sigma} \{F(\sigma)\}^T\, \delta\{D(\sigma)\}\, d\sigma \tag{2.56}$$

in which $\{AD\}$ is understood to represent matrix $[A]$ operating on $\{D\}$.

The kinetic energy is given by

$$\delta K_E = \int_{\mathscr{V}} \rho \{\dot{q}\}^T\, \delta\{\dot{q}\}\, d\mathscr{V} \tag{2.57}$$

or, if the system is at rest at t_0 and t_1, integration by parts gives

$$\delta K_E = -\int_{\mathscr{V}} \rho \{\ddot{q}\}^T\, \delta\{q\}\, d\mathscr{V}. \tag{2.58}$$

Finally, introducing Eqs 2.56 and 2.58 into Eqs 2.54 and 2.53, we have

$$\delta \bar{L} = \int_{t_0}^{t_1} \left[\int_{\mathscr{S}} \{AD\}^T [H] \, \delta[AD] \, d\mathscr{S} - \int_{\mathscr{S}} \{N_T\}^T \, \delta\{AD\} \, d\mathscr{S} \right.$$

$$- \int_{\mathscr{S}} \{f\}^T \, \delta\{D\} \, d\mathscr{S} - \int_{\sigma} \{F(\sigma)\}^T \, \delta\{D(\sigma)\} \, d\sigma \quad (2.59)$$

$$\left. + \int_{V} \rho \{\ddot{q}\}^T \, \delta\{q\} \, d\mathscr{V} \right] dt = 0$$

Equation 2.59 may be written in harmonic form as well. Also, the influence of viscous damping may be introduced.

2.7.2 Principle of stationary total potential energy

For a static problem, the integral over time is superfluous and the last term of Eq. 2.59 is zero. The remaining terms are associated with the total potential energy functional G and constitute the well-known Principle of Stationary Total Potential Energy (PSPE); $\delta G = 0$:

$$\delta G = \int_{\mathscr{S}} \{AD\}^T [H] \, \delta\{AD\} \, d\mathscr{S} - \int_{\mathscr{S}} \{N_T\}^T \, \delta\{AD\} \, d\mathscr{S}$$

$$- \int_{\mathscr{S}} \{f\}^T \, \delta\{D\} \, d\mathscr{S} - \int_{\sigma} \{F(\sigma)\}^T \, \delta\{D(\sigma)\} \, d\sigma = 0 \quad (2.60)$$

In the PSPE, it is understood that only variations of the displacements are permitted.

2.7.3 Reissner's principle

It is sometimes advantageous to retain explicit variables other than displacements $\{D\}$ in the formulation. One possibility is the stress resultants and stress couples $\{N\}$, another is the strains and changes in curvatures $\{\varepsilon\}$. However, neither $\{N\}$ nor $\{\varepsilon\}$ have been shown to be particularly beneficial when compared with $\{D\}$. But, a less obvious alternative combining parts of $\{D\}$ and $\{N\}$ has been shown to have some advantages. Since the matrix of dependent variables is a combination, the resulting formulation is properly called a *mixed* finite element method. Here we consider only the isotropic static case, but the extension to the dynamic problem is straightforward.

E. Reissner derived a general variational theorem which is suitable for our purposes[10] (RVP). The theorem is based on the functional I which incorporates the kinematic law, Eq. 2.30, and states that the governing equations of thin shell theory can be derived from applying the stationary condition to I, i.e. from $\delta I = 0$. Prato obtained a contracted form of the general variational principle by identically satisfying the isotropic, isothermal constitutive law; the moment equilibrium equations (which are known from other formulations of the problem); and certain stress and displacement boundary conditions.[11] There, the loading was restricted to forces (no moments).

For this case, the functional I takes the form

$$I = \int_{\mathscr{S}} [W_M - W_B + U + Q_\phi(\gamma_\phi - \beta_\phi) + Q_\theta(\gamma_\theta - \beta_\theta)] \, d\mathscr{S}$$

$$+ \sum_{\mathscr{S}} \int_\sigma \{F_1(\sigma)\}^T \{D_1(\sigma)\} \, d\sigma - \int_{\sigma_b} \{F_1(\sigma_b)\}^T \{D_1(\sigma_b)\} \, d\sigma_b$$

$$- \int_{\nu_b} \{F_2(\nu_b)\}^T \{D_2(\nu_b)\} \, d\nu_b \quad (2.61)$$

in which

$$W_M = [Eh/2(1-\mu^2)][\varepsilon_\phi^2 + \varepsilon_\theta^2 + 2\mu\varepsilon_\phi\varepsilon_\theta + 2(1-\mu)\varepsilon_{\theta\phi}^2] \quad (2.62)$$
= membrane strain energy density

$$W_B = [6/Eh^3][M_\phi^2 + M_\theta^2 - 2\mu M_\phi M_\theta + 2(1+\mu)M_{\theta\phi}^2] \quad (2.63)$$
$$+ [(1+\mu)/\lambda Eh][Q_\phi^2 + Q_\theta^2]$$
= complementary strain energy density associated with bending and transverse shear deformation

$$U = \{f_1\}^T \{D_1\} \quad (2.64)$$

Also,

$$\{f_1\} = \{f_u f_v f_w\} \quad (2.65)$$
$$\{F_1\} = \{F_u F_v F_w\} \quad (2.66)$$
$$\{F_2\} = \{F_{\beta_\phi} F_{\beta_\theta}\} \quad (2.67)$$

and

$$\{D_1\} = \{u \, v \, w\} \quad (2.68)$$
$$\{D_2\} = \{\beta_\phi \beta_\theta\} \quad (2.69)$$

Furthermore, $\{F_1(\sigma_b)\}$ are prescribed forces along boundary σ_b and $\{D_2(\nu_b)\}$ are prescribed rotations along boundary ν_b.

The stationary condition follows from

$$\delta I = 0 \quad (2.70)$$

after the functional has been written in terms of u, v, w, M_ϕ, M_θ and $M_{\theta\phi}$. This is accomplished by utilizing Eq. 2.30 to eliminate the strains, and the aforementioned moment equilibrium equations to eliminate the transverse shear resultants. For a rotational shell, the latter equations take the harmonic form

$$Q_\phi = (\cos \phi/R)(M_\phi - M_\theta) + (1/R_\phi)M_{\phi,\phi} + (1/R)M_{\theta\phi,\theta} \qquad (2.71)$$

$$Q_\theta = (2 \cos \phi/R)M_{\theta\phi} + (1/R_\phi)M_{\theta\phi,\phi} + (1/R)M_{\theta,\theta}. \qquad (2.72)$$

2.7.4 Solution of variational problems

The variational problems embodied in the previously stated variational principles are conveniently solved by using the *direct* method of the Calculus of Variations, whereby weighted comparison functions are chosen to represent the explicit dependent variables such as $\{D\}$ in Eqs 2.59 and 2.60. Such functions should also satisfy the appropriate boundary conditions. The weighting factors assigned to the comparison functions are then determined from the stationary condition of the functional. In effect, the problem is transformed to a maximum–minimum problem of classical calculus and a set of simultaneous algebraic equations for the weighting factors is produced. This procedure is widely employed and is known as the Rayleigh–Ritz method.[12]

The *global* stationary problem, as indicated by Eqs 2.59 and 2.70, is assumed to be represented by the assemblage of elements which have satisfied the *local* stationary condition, since the functional for the complete system is comprised of the sum of the functionals of the individual regions or elements. Furthermore, the energy principles can be applied to individual elements disregarding the interelement boundary conditions, provided the element is compatible[13] which will be the case in this treatment. Thus, the comparison functions need be chosen only over the *element domain* with consideration of the physical boundary conditions deferred to the global level assemblage of element equations.

The possibility of choosing comparison functions over subdomains or elements and needing only to satisfy conditions at nodes incident to the element is perhaps the main appeal of a finite element formulation. This is illustrated in the following chapters.

References

1. Gould, P. L., *Static Analysis of Shell Structures*, Lexington Press, D. C. Heath and Co., 1977.
2. Brombolich, L. J. and Gould, P. L., 'Finite Element Analysis of Shells of Revolution by Minimization of the Potential Energy Functional', *Proc. of the Symposium on Applications of the Finite Elements in Civil Engineering*, Vanderbilt Univ., Nashville, Tenn., 1969, pp. 279–307.

3. Lukasiewicz, S., *Local Loads in Plates and Shells*, Sijthoff & Noordhoff, The Netherlands, 1979.
4. Bushnell, D., 'Buckling of Shells—Pitfall for Designers', *AIAA J.*, Vol. 19, No. 9, Sept. 1981, pp. 1183–1226.
5. Gould, P. L. and Basu, P. K., 'Geometrical Stiffness Matrices for the Finite Element Analysis of Rotational Shells', *J. Struct. Mech.*, Vol. 5, No. 1, 1977, pp. 87–105.
6. Bushnell, D., 'Computerized Analysis of Shells-Governing Equations', *J. Computers and Structures*, Vol. 18, No. 3, 1984, pp. 471–536.
7. Basu, P. K. and Gould, P. L., 'Dynamic and Nonlinear Analysis of Shells of Revolution', Research Report No. 46, Structural Division, Dept. of Civil Engineering, Washington University, St. Louis, Mo., April 1977.
8. Novozhilov, V. V., *Thin Shell Theory* (Translated from 2nd Russian ed. by P. G. Lowe), Noordhoff, Groningen, The Netherlands, 1964, pp. 151–163.
9. Sen, S. K. and Gould, P. L., 'Free Vibration of Shells Using FEM', *J. Engrg. Mech. Div., ASCE*, Vol. 100, No. EM2, April 1974, pp. 283–303.
10. Reissner, E., 'On a Variational Theorem in Elasticity', *J. of Math. and Physics*, Vol. 29, 1950, pp. 90–95.
11. Prato, C. A., 'Shell Finite Element Method Via Reissner's Principle', *Int. J. Solids and Structures*, Vol. 5, 1969, pp. 1119–1133.
12. Forray, M. J., *Variational Calculus in Science and Engineering*, McGraw-Hill Book Co., New York, 1968, pp. 157–188.
13. Gallagher, R. H., *Finite Element Analysis Fundamentals*, Prentice-Hall Inc., Englewood Cliffs, NJ, 1975, pp. 152–156.

3 Static analysis

3.1 Representation of dependent variables

3.1.1 Comparison functions

A key step in the finite element approach is to represent the dependent variables which appear explicitly in the formulation by comparison functions over the element domain. In the case of rotational shells, we need be concerned with only the meridional variation since harmonic analysis is applied in the circumferential direction.

Consider a typical variable x^j, which could be an element of $\{D^j\}$ (Eq. 2.19) in the case of a displacement formulation, or an element of $\{N^j\}$ (Eq. 2.24) in the case of a force or mixed formulation. Within a finite element i, we take

$$x^j = \lfloor Z_x \rfloor \{Y_x^j\} \tag{3.1}$$

in which

$$\lfloor Z_x \rfloor = \lfloor z_{x1} z_{x2} -- z_{xm} -- z_{xn} \rfloor \tag{3.2}$$

$$\{Y_x^j\} = \{y_{x1} y_{x2} -- y_{xm} -- y_{xn}\} \tag{3.3}$$

where the elements of $\lfloor Z_x \rfloor$ are *shape functions* and the elements of $\{Y_x^j\}$ are *constant coefficients*.

This form of approximation is quite flexible in that over element i and harmonic j, each variable x^j may be represented by a polynomial of order n. This produces n unknown constant coefficients y_{xm} ($m=1,\ldots,n$) which need to be evaluated to weigh each shape function z_{xm}. For bookkeeping purposes, it is often convenient to limit this flexibility by using the same order of approximation for each variable, element and/or harmonic; this is not necessary, but will be followed here.

Consider now the detailed form of the shape function

$$\lfloor Z_x \rfloor = \lfloor z_{x1} z_{x2} z_{x3} \ldots z_{xm} \ldots z_{xn} \rfloor \tag{3.4}$$

In terms of the meridional variable $s = s_i$ as defined by Eq. 2.7, $z_{x1} = (1-s)$, $z_{x2} = s$ and $z_{x3} = s(1-s)$ are very suitable. For $m > 3$, there are a number of possibilities. Two which have merit are

$$z_{xm} = s^{m-2}(1-s) \tag{3.5a}$$

and

$$z_{xm} = s^{m-2}(1-s) - s(s-1)^{m-2} \tag{3.5b}$$

With respect to the constant coefficients, it is first noted that the resulting polynomials of degree n should satisfy C^0 continuity (C^0 indicates the 'zeroth' order derivative, or the variable itself) at the interelement boundaries. Coefficient y_{x1} will be equal to the nodal variable x^j at $s = 0$ and coefficient y_{x2} will be equal to x^j at $s = 1$ since the higher order terms $m \geq 2$ vanish at the end points. The coefficients y_{xm} ($m > 2$) represent internal nodal variables which need not be physically interpreted; the associated modes z_{xm} ($m > 2$) are so-called 'bubble' functions which serve to refine the approximation at the element level but which may be condensed out before proceeding to the global level.

It is appropriate to note here that C^0 continuity is a desirable characteristic, made possible by retaining transverse shearing strains. The minimum continuity required at interelement boundaries is related to the highest order derivative appearing in the strain energy density function, initially the first term on the r.h.s. of Eq. 2.51. It has been shown that if the highest order derivative is the first, then only the nodal variables themselves must be continuous.[1] An examination of the several variational expressions in Section 2.7, with the constitutive laws used to write stresses in terms of strains and the kinematic law used to convert the entire expression to displacements, reveals no derivatives beyond the first, so that C^0 is indeed sufficient. Had transverse shearing strains been suppressed, the kinematic law would contain second derivatives of displacements and C^1 continuity would be required, a less advantageous situation.[1]

The hierarchic property of these shape functions, as discussed by Basu,[2] indicates that an interpolation of degree n constitutes a subset of the set of shape functions corresponding to an interpolation of degree $n + 1$. The same will hold true for the stiffness and mass matrices, which are developed subsequently, for the element of order n as related to the element of order $n + 1$. This property has been successfully exploited in several second generation finite element codes but has not been incorporated in the numerical studies reported in this book.

The alternative forms of the higher order shape functions represented by Eqs 3.5(a) and (b) have also been studied by Basu.[2] The first form, Eq. 3.5(a) has proved to be satisfactory in a wide variety of numerical studies recorded in the literature; however, the other form, Eq. 3.5(b), has been shown to possess symmetrical and antisymmetrical properties which are useful for explicitly identifying the corresponding coefficients y_{xm} ($m > 2$),

Finite element analysis of shells of revolution

and for improving the convergence. Only very limited numerical comparisons are available, but the indications are that new generation codes should consider employing the later type of shape functions. Again, the numerical studies reported in this book do *not* include this development.

3.1.2 Approximations at element level

We consider the two formulations which are relevant to this book, the displacement formulation based on the PSPE and the mixed formulation based on the RVP. For the first case, the dependent variables for element i are displacements defined by Eqs 2.17 and 2.19 and approximated as

$$\{D_{ij}^i\} = \lceil \Theta_1^i \rfloor [Z_i]\{Y_{ij}^i\} \tag{3.6}$$

in which

$$[Z_i] = \begin{bmatrix} \lfloor Z_u \rfloor & & & & \\ & \lfloor Z_v \rfloor & & & \\ & & \lfloor Z_w \rfloor & & \\ & & & \lfloor Z_{\beta_\phi} \rfloor & \\ & & & & \lfloor Z_{\beta_\theta} \rfloor \end{bmatrix} \tag{3.7}$$

$$\{Y_{ij}^i\} = \{Y_u^i Y_v^i Y_w^i Y_{\beta_\phi}^i Y_{\beta_\theta}^i\} \tag{3.8}$$

and $\lceil \theta_1^i \rfloor$ is given by Eq. 2.18.

A typical interpolation function is

$$\lfloor Z_u \rfloor = \lfloor (1-s)s \; s(1-s) \; s^2(1-s) \ldots \rfloor$$

For the mixed formulation, the vector of dependent variables is

$$\{M^i\} = \{u^i v^i w^i M_\phi^i M_\theta^i M_{\theta\phi}^i\} \tag{3.9}$$

The corresponding approximations will be

$$\{M_{ij}^i\} = \lceil \Theta_5^i \rfloor [Z_i]\{Y_{ij}^i\} \tag{3.10}$$

in which

$$[Z_i] = \begin{bmatrix} \lfloor Z_u \rfloor & & & & & \\ & \lfloor Z_v \rfloor & & & & \\ & & \lfloor Z_w \rfloor & & & \\ & & & \lfloor Z_{m_\phi} \rfloor & & \\ & & & & \lfloor Z_{m_\theta} \rfloor & \\ & & & & & \lfloor Z_{m_{\theta\phi}} \rfloor \end{bmatrix} \tag{3.11}$$

$$\{Y_{ij}^i\} = \{Y_u^i Y_v^i Y_w^i Y_{m_\phi}^i Y_{m_\phi}^i Y_{m_{\theta\phi}}^i\} \tag{3.12}$$
$$= \text{vector of constant coefficients}$$

and
$$\lceil \Theta_5^i \rfloor = \lceil \cos j\theta \ \sin j\theta \ \cos j\theta \ \cos j\theta \ \cos j\theta \ \sin j\theta \rfloor \tag{3.13}$$

These approximations are utilized in the subsequent sections where the displacement and mixed finite element models are developed.

3.2 Linear displacement formulation

3.2.1 Element equilibrium equation

We refer to the PSPE, as discussed in Section 2.7.2 with the displacement functions given by Eq. 3.6, and consider the nodal variables on the interval $(0 \le s_i \le 1)$:

$$\{\Delta_i^j\} = \{u_i^j(0) v_i^j(0) w_i^j(0) \beta_{\phi i}^j(0) \beta_{\theta i}^j(0) u_i^j(1) v_i^j(1) w_i^j(1) \beta_{\phi i}^j(1) \beta_{\theta i}^j(1)\} \tag{3.14}$$

We rearrange the coefficients $\{Y_{ij}^i\}$, as defined by Eq. 3.8, into those terms which are nodal variables and those which are not.

$$\{\bar{Y}_{ij}^i\} = \{y_{u1} y_{v1} y_{w1} y_{\beta_\phi 1} y_{\beta_\theta 1} y_{u2} y_{v2} y_{w2} y_{\beta_\phi 2} y_{\beta_\theta 2} :$$
$$y_{u3} \cdots y_{un} y_{v3} \cdots y_{vn} y_{w3} \cdots y_{wn} y_{\beta_\phi 3} \cdots y_{\beta_\phi n} y_{\beta_\theta 3} \cdots y_{\beta_\theta n}\}$$
$$= \{\bar{Y}_{ia}^j : \bar{Y}_{ib}^j\} \tag{3.15}$$

From the discussion in Section 3.1.1, it is apparent that

$$\{\bar{Y}_{ia}^j\} = \{\Delta_i^j\} \tag{3.16}$$

so that

$$\{\bar{Y}_i^j\} = \{\Delta_i^j : \bar{Y}_{ib}^j\} \tag{3.17}$$

which may be called the vector of *generalized displacements*. It is then convenient to rewrite Eq. 3.6 as

$$\{D_i^j\} - \lceil \Theta_1 \rfloor \lceil \bar{Z}_i \rceil \{\bar{Y}_i^j\} \tag{3.18}$$

in which
$$[\bar{Z}_i] = [\bar{Z}_{ia} : \bar{Z}_{ib}] \tag{3.19}$$

$$[\bar{Z}_{ia}] = \begin{bmatrix} 1-s & 0 & 0 & 0 & 0 & s & 0 & 0 & 0 & 0 \\ 0 & 1-s & 0 & 0 & 0 & 0 & s & 0 & 0 & 0 \\ 0 & 0 & 1-s & 0 & 0 & 0 & 0 & s & 0 & 0 \\ 0 & 0 & 0 & 1-s & 0 & 0 & 0 & 0 & s & 0 \\ 0 & 0 & 0 & 0 & 1-s & 0 & 0 & 0 & 0 & s \end{bmatrix} \tag{3.20}$$

Finite element analysis of shells of revolution

$$[\bar{Z}_{ib}] = \begin{bmatrix} \lfloor Z_b \rfloor & 0 & 0 & 0 & 0 \\ 0 & \lfloor Z_b \rfloor & 0 & 0 & 0 \\ 0 & 0 & \lfloor Z_b \rfloor & 0 & 0 \\ 0 & 0 & 0 & \lfloor Z_b \rfloor & 0 \\ 0 & 0 & 0 & 0 & \lfloor Z_b \rfloor \end{bmatrix} \quad (3.21a)$$

and
$$\lfloor Z_b \rfloor = \lfloor s(1-s) \; s^2(1-s) \ldots s^{n-2}(1-s) \rfloor \quad (3.21b)$$

Now, we evaluate δG, as defined by Eq. 2.60, for element i and harmonic j by substituting Eq. 2.17 for $\{D\}$ in view of Eq. 3.18; by representing $D(\sigma)$ in the same way, with $\{D^j(s)\}$ taken as the nodal values $\{\Delta_i^j\}$; and by taking $[A]$ as Eq. 2.31 for a general rotational element, or as Eq. 2.33 for a closed element.

The known force terms are represented in the appropriate harmonic forms as given by Eqs 2.23–2.28. It is recognized that the boundaries are along nodal circles and that the ring loads are combined with the boundary forces in the last term of Eq. 2.60. For rearrangement manipulations, the matrix multiplication rule $[CD]^T = [D]^T[C]^T$ is applied repeatedly, and it is also assumed that the variations $\delta\{\bar{Y}_i^j\}$ are arbitrary and independent of s and θ. The result is

$$\delta G_i^j = \delta\{\bar{Y}_i^j\}^T \bigg[\int_{\mathscr{S}} [B]^T [H][B]\{\bar{Y}_i^j\} \, d\mathscr{S}$$
$$- \int_{\mathscr{S}} [B]^T \lceil \Theta_2^i \rceil \{N_{T_i}^i\} \, d\mathscr{S}$$
$$- \int_{\mathscr{S}} [\bar{B}]^T \lceil \Theta_1^i \rceil \{f_i^j\} \, d\mathscr{S}$$
$$- \int_{\theta} \lceil \Theta_1^i \rceil \lceil \Theta_1^i \rceil \{F_i^j\} R \, d\theta \bigg] = 0 \quad (3.22)$$

in which
$$[B] = [A^j][\Theta_1^j][\bar{Z}_i] \quad (3.23)$$
$$[\bar{B}] = \lceil \Theta_1^i \rceil [\bar{Z}_i] \quad (3.23b)$$

and the incremental arc length along the circumference is
$$d\sigma = R \, d\theta \quad (3.23c)$$

It should also be recognized that the matrix $[B]^T[H][B]$ is symmetric since $[H]$ is symmetric; this property was used in the derivation of Eq. 3.22.

Several simplifications are in order at this point. First, the differential surface area is

$$d\mathscr{S} = RL_i \, d\theta \, ds \qquad (3.24)$$

in accordance with Eqs 2.1–2.7. Next, only the terms with a j superscript are harmonic dependent; only the Θ^j matrices are dependent on the circumferential coordinates; and the Fourier coefficients are not functions of the surface variables. This leads to the following collected terms:

$$[k_i^j] = L_i \int_{-\pi}^{\pi} \int_0^1 [B]^T[H][B]R \, ds \, d\theta$$

$$= L_i \lambda \pi \int_0^1 [B]^T[H][B]R \, ds \qquad (3.25)$$

= element stiffness matrix

The constant $\lambda = 2$ for $j = 0$ and $\lambda = 1$ for $j \geq 1$

$$\{\bar{N}_{Ti}^j\} = L_i \lambda \pi \int_0^1 [B]^T \{N_{Ti}^j\} R \, ds \qquad (3.26)$$

= element thermal load vector

$$\{\bar{f}_i^j\} = L_i \lambda \pi \int_0^1 [\bar{B}]^T \{f_i^j\} R \, ds \qquad (3.27)$$

= element distributed load vector

$$\{\bar{F}_i^j\} = \lambda \pi R \{F_i^j\} \qquad (3.28)$$

= element ring load vector

Because the load vectors given by Eqs 3.26–3.28 are derived with the same interpolations used for the displacements, they are called (variationally) *consistent load vectors*. Also, because the ring loads are directly applied to the nodes, there are no terms in $\{\bar{F}_i^j\}$ corresponding to the $\{\bar{Y}_{ib}^j\}$ vector in Eq. 3.17.

Since $\delta\{\bar{Y}_i^j\}^T$ is arbitrary, Eq. 3.22 is satisfied by setting the bracketed term equal to zero which gives

$$[k_i^j]\{\bar{Y}_i^j\} = \{N_{Ti}^j\} + \{\bar{f}_i^j\} + \{\bar{F}_i^j\} = \{\mathscr{F}_i^j\} \qquad (3.29)$$

as the equilibrium equation for element i and harmonic j.

3.2.2 Condensed equilibrium equation

It is of interest to examine the vector of generalized displacements $\{\bar{Y}_i^j\}$, as given by Eq. 3.17, more closely. Since the second part of the vector $\{\bar{Y}_{ib}^j\}$ is composed of intra-element variables only, it is advantageous to remove them from the equilibrium equations prior to proceeding to the global level. The transformation is known as *static condensation* and is implemented by writing Eq. 3.29 in the partitioned form

$$\begin{bmatrix} \mathbf{k}_{iaa}^j & \mathbf{k}_{iab}^j \\ \mathbf{k}_{iba}^j & \mathbf{k}_{ibb}^j \end{bmatrix} \begin{Bmatrix} \mathbf{\Delta}_i^j \\ \mathbf{\bar{Y}}_{ib}^j \end{Bmatrix} = \begin{Bmatrix} \mathbf{\bar{N}}_{Tia} \\ \mathbf{\bar{N}}_{Tib} \end{Bmatrix} + \begin{Bmatrix} \mathbf{\bar{f}}_{ia}^j \\ \mathbf{\bar{f}}_{ib}^j \end{Bmatrix} + \begin{Bmatrix} \mathbf{\bar{F}}_i^j \\ 0 \end{Bmatrix} \quad (3.30)$$

Equation 3.30 is then expanded to

$$\mathbf{k}_{iaa}^j \mathbf{\Delta}_i^j + \mathbf{k}_{iab}^j \mathbf{\bar{Y}}_{ib}^j = \mathbf{\bar{N}}_{Tia} + \mathbf{\bar{f}}_{ia}^j + \mathbf{\bar{F}}_i^j \quad (3.31a)$$

$$\mathbf{k}_{iba}^j \mathbf{\Delta}_i^j + \mathbf{k}_{ibb}^j \mathbf{\bar{Y}}_{ib}^j = \mathbf{\bar{N}}_{Tib} + \mathbf{\bar{f}}_{ib}^j + 0 \quad (3.31b)$$

Solving Eq. 3.31(b) for $\mathbf{\bar{Y}}_{ib}^j$ gives

$$\mathbf{\bar{Y}}_{ib}^j = \mathbf{k}_{ibb}^{j-1} \{ \mathbf{\bar{N}}_{Tib} + \mathbf{f}_{ib} - \mathbf{k}_{iba}^j \mathbf{\Delta}_i^j \} \quad (3.32)$$

and substituting into Eq. 3.31(a) yields

$$[\mathbf{\bar{k}}_i^j]\{\mathbf{\Delta}_i^j\} = \{\mathbf{\bar{\mathcal{F}}}_i^j\} \quad (3.33)$$

in which

$$[\mathbf{\bar{k}}_i^j] = [\mathbf{k}_{iaa}^j - \mathbf{k}_{iab}^j \mathbf{k}_{ibb}^{j-1} \mathbf{k}_{iba}^j] \quad (3.34)$$
$$= \text{condensed element stiffness matrix}$$

and

$$\{\mathbf{\bar{\mathcal{F}}}_i^j\} = \{\mathbf{\bar{N}}_{Tia} + \mathbf{\bar{f}}_{ia}^j + \mathbf{\bar{F}}_i^j - \mathbf{k}_{iab}^j \mathbf{k}_{ibb}^{j-1}(\mathbf{\bar{N}}_{Tib} + \mathbf{\bar{f}}_{ib}^j)\}$$
$$= \{\mathbf{\bar{\mathcal{F}}}_i^j(0) \, \mathbf{\bar{\mathcal{F}}}_i^j(1)\} \quad (3.35)$$
$$= \text{element nodal force vector}$$

The condensed element stiffness matrix $[\mathbf{\bar{k}}_i^j]$ will be used in all global level computations. Vector $\{\mathbf{\bar{\mathcal{F}}}_i^j\}$ gives the contributions of the loads acting on element i to the equivalent nodal forces at nodes $i-1$ and i at the global level. Once the nodal displacements have been evaluated, the condensed variables $\mathbf{\bar{Y}}_{ib}^j$, which are useful for intra-element interpolations using Eq. 3.17, may be restituted through Eq. 3.32.

3.2.3 Global equilibrium equations

3.2.3.1 Assembly

At first, it is assumed that the shell meridian is smooth with no slope discontinuities. Recognizing that Eq. 3.33 may be partitioned as

$$\begin{bmatrix} \mathbf{\bar{k}}_{iaa}^j & \mathbf{\bar{k}}_{iab}^j \\ \mathbf{\bar{k}}_{iba}^j & \mathbf{\bar{k}}_{ibb}^j \end{bmatrix} \begin{Bmatrix} \mathbf{\Delta}_{ia}^j \\ \mathbf{\Delta}_{ib}^j \end{Bmatrix} = \begin{Bmatrix} \mathbf{\bar{\mathcal{F}}}_{ia}^j \\ \mathbf{\bar{\mathcal{F}}}_{ib}^j \end{Bmatrix} \quad (3.36)$$

and referring to Eq. 3.15, it is obvious that $\mathbf{\Delta}_{ib}^j$, the nodal displacements at $s=1$ for element i, are equal to $\mathbf{\Delta}_{i+1,a}^j$, the nodal displacements at $s=0$ for element $i+1$. This leads to the assembly of the global equilibrium equations using the *direct stiffness method*, which takes the form

$$\begin{bmatrix} \bar{k}^j_{1aa} & \bar{k}^j_{1ab} & & \\ \bar{k}^j_{1ba} & \bar{k}^j_{1bb}+\bar{k}^j_{2aa} & \bar{k}^j_{2ab} & \\ & \bar{k}^j_{2ba} & \bar{k}^j_{2bb}+\bar{k}^j_{3aa} & \\ & & \vdots & \end{bmatrix} \begin{Bmatrix} \Delta^j_{1a} \\ \Delta^j_{1b}=\Delta^j_{2a} \\ \Delta^j_{2b}=\Delta^j_{3a} \\ \vdots \end{Bmatrix} = \begin{Bmatrix} \bar{\mathcal{F}}^j_{1a} \\ \bar{\mathcal{F}}^j_{1b}+\bar{\mathcal{F}}^j_{2a} \\ \bar{\mathcal{F}}^j_{2b}+\bar{\mathcal{F}}^j_{3a} \\ \vdots \end{Bmatrix} \quad (3.37)$$

For a model with N elements, this produces $5N$ simultaneous equations for $j \geq 1$ and $3N$ for $j = 0$.

3.2.3.2 Boundary conditions

The next step is to apply the displacement boundary conditions, which are taken to be homogeneous. Several possibilities were presented in Section 2.6. Computationally, this is achieved by eliminating the rows and columns of Eq. 3.37 corresponding to the 'zeroed' variables, ultimately producing

$$[\bar{K}^j]\{\Delta^j\} = \{\bar{\mathcal{F}}^j\} \quad (3.38)$$

in which $[\bar{K}^j]$ is the global stiffness matrix.

In this regard, it should be noted that sufficient kinematic conditions to eliminate rigid body motion should be enforced. Technically, the following procedure is only approximate because of the nature of curvilinear coordinates but it has been shown to be adequate whenever applied: For $j = 0$, it is sufficient to eliminate u at one node; for $j = 1$, u and v must both be eliminated at one node (with some exceptions), while for $j > 1$, the loading is self-equilibrated and no kinematic constraints are required. For a completely closed shell, such as a pressurized sphere, it will be necessary to enforce symmetry conditions to utilize this formulation. Failure to eliminate rigid body motion will result in a singular global stiffness matrix. This is discussed further in Section 3.2.3.5.

3.2.3.3 Solution

It is obvious from Eq. 3.37 that the global equations are banded; therefore, they may be solved efficiently by a variety of methods, such as Gaussian elimination for banded equations.[3] It is convenient to write the solution symbolically as

$$\{\Delta^j\} = [\bar{K}^j]^{-1}\{\bar{\mathcal{F}}^j\} \quad (3.39)$$

even though $[\bar{K}^j]$ may not actually be inverted in the solution algorithm.

The element nodal displacement vectors may be obtained from $\{\Delta^j\}$ after the 'zeroed' elements have been reinserted. Finally, the eliminated variables $\{\bar{Y}^j_{ib}\}$ are evaluated by Eq. 3.32, whereupon the full high-order approximation for the displacements in element i, $\{\bar{Y}^j_i\}$, will be available from Eq. 3.17.

3.2.3.4 Strains and stress resultants

In order to compute the strains and stress resultants, it is convenient to return from $\{\bar{Y}_i^j\}$ to $\{Y_i^j\}$, as defined by Eq. 3.12. Then, the displacement vector $\{D_i^j\}$ is found from Eq. 3.6 as a continuous function of s. The presence of the higher order terms markedly improves the quality of the approximation at this stage. This becomes apparent when we refer to the kinematic law Eq. 2.30, where differentiations are required. Thus, at least a quadratic approximation for each displacement variable is required to provide a linear variation in strain. This would seem to be the minimum acceptable and is usually too crude.

Computationally, for a general rotational element, the strains are found by applying Eq. 2.30 to $\{D_i^j\}$ as given by Eq. 3.6:

$$\{\varepsilon_i\}^j = \lceil\Theta_2^j\rfloor\{\varepsilon_i^j\} = \lceil\Theta_2^j\rfloor[\bar{A}_i^j]\{Y_i^j\} \tag{3.40}$$

in which $\lceil\Theta_2^j\rfloor$ is defined in Eq. 2.21 and

$[\bar{A}_i^j] =$

$$\begin{bmatrix} (1/L_i)\lfloor Z_u\rfloor_{,s} & 0 & (1/R_\phi)\lfloor Z_w\rfloor & 0 & 0 \\ \left(\dfrac{\cos\phi}{R}\right)\lfloor Z_u\rfloor & \left(\dfrac{j}{R}\right)\lfloor Z_v\rfloor & \left(\dfrac{\sin\phi}{R}\right)\lfloor Z_w\rfloor & 0 & 0 \\ \left(\dfrac{-j}{R}\right)\lfloor Z_u\rfloor & \begin{bmatrix}\left(\dfrac{1}{L_i}\right)\lfloor Z_v\rfloor_{,s} \\ -\left(\dfrac{\cos\phi}{R}\right)\lfloor Z_v\rfloor\end{bmatrix} & 0 & 0 & 0 \\ 0 & 0 & 0 & \left(\dfrac{1}{L_i}\right)\lfloor Z_{\beta_\phi}\rfloor_{,s} & 0 \\ 0 & 0 & 0 & \left(\dfrac{\cos\phi}{R}\right)\lfloor Z_{\beta_\phi}\rfloor & \left(\dfrac{j}{R}\right)\lfloor Z_{\beta_\theta}\rfloor_{,s} \\ 0 & 0 & 0 & \left(\dfrac{-j}{2R}\right)\lfloor Z_{\beta_\phi}\rfloor & \begin{bmatrix}(1/2L_i)\lfloor Z_{\beta_\theta}\rfloor_{,s} \\ -\left(\dfrac{\cos\phi}{2R}\right)\lfloor Z_{\beta_\theta}\rfloor\end{bmatrix} \\ (-1/R_\phi)\lfloor Z_u\rfloor & 0 & (1/L_i)\lfloor Z_w\rfloor_{,s} & 1 & 0 \\ 0 & \left(\dfrac{-\sin\phi}{R}\right)\lfloor Z_v\rfloor & \left(\dfrac{-j}{R}\right)\lfloor Z_w\rfloor & 0 & 1 \end{bmatrix}$$

$$\tag{3.41}$$

Similarly for a cap element $(i = 0)$, the strains are found by applying Eq. 2.32 to $\{D_0^j\}$:

$$\{\varepsilon_i\}^j = \lceil\Theta_2^j\rfloor\{\varepsilon_0^j\} = \lceil\Theta_2^j\rfloor[\bar{A}_0^j]\{Y_0^j\} \tag{3.42}$$

in which

$[\bar{A}_0^i] =$

$$\begin{bmatrix} (1/L_0)\lfloor Z_u \rfloor_{,s} & 0 & [1/R_\phi(0)]\lfloor Z_w \rfloor & 0 & 0 \\ (1/L_0)\lfloor Z_u \rfloor_{,s} & (j/L_0)\lfloor Z_v \rfloor_{,s} & [1/R_\phi(0)]\lfloor Z_w \rfloor & 0 & 0 \\ (-j/L_0)\lfloor Z_u \rfloor_{,s} & 0 & 0 & 0 & 0 \\ 0 & 0 & 0 & (1/L_0)\lfloor Z_{\beta_\phi} \rfloor_{,s} & 0 \\ 0 & 0 & 0 & (1/L_0)\lfloor Z_{\beta_\phi} \rfloor_{,s} & (j/L_0)\lfloor Z_{\beta_\theta} \rfloor_{,s} \\ 0 & 0 & 0 & (-j/2L_0)\lfloor Z_{\beta_\phi} \rfloor_{,s} & 0 \\ [-1/R_\phi(0)]\lfloor Z_u \rfloor & 0 & (1/L_0)\lfloor Z_w \rfloor_{,s} & \lfloor Z_{\beta_\phi} \rfloor & 0 \\ 0 & [-1/R_\phi(0)]\lfloor Z_v \rfloor & (-j/L_0)\lfloor Z_w \rfloor_{,s} & 0 & \lfloor Z_{\beta_\theta} \rfloor \end{bmatrix}$$

(3.43)

Finally, the stress resultants and couples may be evaluated from Eq. 2.46 as

$$\{N_i\}^j = \lceil \Theta_2^i \rfloor \{N_i^j\} = \lceil \Theta_2^i \rfloor [[H]\{\varepsilon_i^j\} - \{N_{Ti}^j\}]$$ (3.44)

where all terms have previously been defined.

It should be noted that both the strains and the stresses, as evaluated in the preceding equations, are continuous functions of s and thus may be calculated at designated locations between nodal circles as well as at the global nodal locations. This permits a quite detailed determination of the local distribution of the dependent variables, even though the global grid may be relatively coarse.

3.2.3.5 Rigid body displacements

Although the formulation of the rotational shell element in curvilinear coordinates has many obvious advantages, it is difficult to provide displacement functions which explicitly ensure an absolutely stress free state under an imposed rigid body displacement. This is beyond the prevailing requirement of providing sufficient constraints to eliminate rigid body motion in loaded shells as discussed in Section 3.2.3.2.

There are two obvious approaches to the resolution of this difficulty. The first is to include *explicit* expressions for some or all of the rigid body terms in the displacement function. This has been implemented, but apparently with limited success.[4] The other is to rely on the precision of the high order displacement functions to *implicitly* represent such effects. Cubic polynomials have been shown to be adequate by Mebane and Stricklin[5] and their approach has been accepted in the current formulation.

It is possible to test the adequacy of the rigid body representation for a particular shell model by performing a free vibration analysis, developed in Chapter 4, on the model with the shell boundaries unrestrained. The lowest few eigenvalues should be close to zero (three for $j=0$ corresponding to u, w, β_ϕ and five for $j>0$ corresponding to u, v, w, β_ϕ, β_θ). Beyond this specific test, no general proof of the adequacy of the implicit representation is offered, but neither are counter-examples to this approach known to the author.

3.3 Extensions to basic element

3.3.1 General

Finite element analysis for shells is most needed when complicated geometrical and configurational conditions are encountered. Examples are stiffened shells, compound shells, branched shells and open shells. Technically, some of these situations fall outside the range of axisymmetric shells but they may still be treated approximately within the present finite element formulation. An elaborate treatment of shells with stiffeners is presented in Ref. 6.

3.3.2 Stiffened shells

3.3.2.1 Classification

It is convenient to consider separately two types of stiffeners (1) rings, or circumferential stiffeners; and (2) stringers, or meridional stiffeners.

3.3.2.2 Circumferential stiffeners

In treating circumferentially stiffened shells within the rotational shell formulation, it is important to specify whether the stiffeners are symmetric with respect to the middle surface of the shell. For cases where the stiffeners are symmetric, they can be conveniently represented as additional individual shell elements, while eccentrically placed stiffeners may be treated as branched shells, which will be discussed later, or by employing a coordinate transformation as suggested by Bushnell.[6]

For the case where the stiffeners are symmetric (or the eccentricity is neglected), an element with appropriate section properties may be directly employed. If the stiffener is a uniform plate, Fig. 3.1, the *depth* of the plate d would be used for the *thickness* of the element, h, while the *thickness* of the plate, b, would define the meridional *length* of the (straight) element, L_i.

If the stiffener is a more complicated shape such as an H-section or pair of channels (but still symmetric), Fig. 3.2, the orthotropic stress–

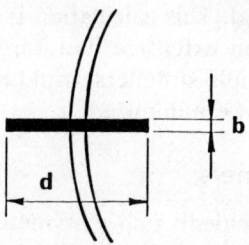

Fig. 3.1 Plate stiffener

strain matrix $[H]$ as defined in Table 2.1 may be used. Following the split-rigidity approach[7] the Poisson's ratio terms should be set to zero making $[H]$ diagonal. All terms except H_{22} and H_{55} and H_{88} should be determined using the basic shell thickness h, while the later terms should be based on equivalent shell thicknesses. For H_{22} and H_{88}, the equivalent in-plane or membrane thickness

$$h_{mi} = \frac{A_s + hL_i}{L_i} \qquad (3.45)$$

where A_s = area of stiffener should be used. For H_{55}, the equivalent bending thickness h_{bi} is found by setting

$$\tfrac{1}{12}L_i h_{bi}^3 = I_s + \tfrac{1}{12}L_i h^3 \qquad (3.46)$$

where I_s = the moment of inertia of the stiffener about an axis tangent to the meridian. Therefore

$$h_{bi} = (12I_s/L_i + h^3)^{1/3} \qquad (3.47)$$

If stiffener is dominant

$$h_{bi} \simeq \left(\frac{12I_s}{L_i}\right)^{1/3} \qquad (3.48)$$

The value of L_i to be used in the preceding calculations should be limited by the 'effective width' of the combined stiffener-shell section

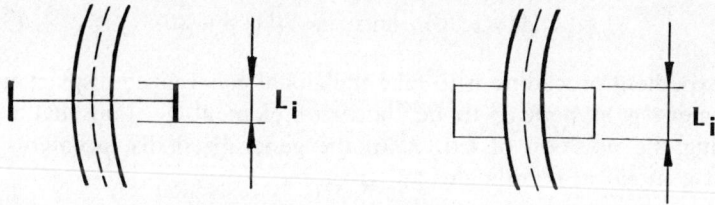

Fig. 3.2 H and channel stiffeners

when shear-lag is considered. This calculation is beyond the scope of the present development but an extensive literature exists on this subject.[8] Ideally, the spacing of multiple stiffeners would be specified such that the loss of effectiveness would be minimized.

3.3.2.3 Meridional stiffeners

Meridional stiffeners are decidedly non-axisymmetric and are thus difficult to include in an axisymmetric shell analysis. Again, regularly spaced or cyclicly symmetrical stiffeners may be approximated by using equivalent shell thicknesses in the orthotropic stress–strain matrix $[H]$ as follows:

$$H_{11} \text{ and } H_{77}: h_m = \frac{A_s + hb}{b} \qquad (3.49)$$

$$H_{44}: h_b = (12I_s/b + h^3)^{1/3} \qquad (3.50)$$

where b = the meridional stiffener spacing as limited by the effective width.

3.3.3 Compound shells

Frequently, shells are composed of segments with different geometries. A familiar example is a closed shell in which the cap may have a different form from the main shell. Also, transitional surfaces are used from a cylinder to a spherical or elliptical cap. Such shells, and others which are defined by more than one meridional curve will be called compound shells. An example is considered in Section 3.6.6.

If the compound shell retains slope continuity, no difficulty is encountered other than a possible slight discrepancy in the calculated stress resultants and couples on either side of the junction. If, however, a slope discontinuity is present, the curvilinear coordinates do not describe unique ϕ and n directions at the junction. This is illustrated on Fig. 3.3.

Considering the displacements of the common junction, node g,

$$v^j_{g-1}(1) = v^j_g(0); \qquad \beta^j_{\phi,g-1}(1) = \beta^j_{\phi,g}(0); \qquad \beta^j_{\theta,g-1}(1) = \beta^j_{\theta,g}(0) \qquad (3.51)$$

but

$$u^j_{g-1}(1) \neq u^j_g(0) \quad \text{and} \quad w^j_{g-1}(1) \neq w^j_g(0) \qquad (3.52)$$

An expedient procedure is to take the global generalized displacements for element g at node g to be those for element $g-1$ at that node. Following the notation of Eq. 3.36, the generalized displacements for element g in *global* coordinates are given by

$$\{\bar{\mathbf{\Delta}}^j_g\} = \{\mathbf{\Delta}^j_{ga}\ \mathbf{\Delta}^j_{gb}\} \qquad (3.53)$$

Static analysis

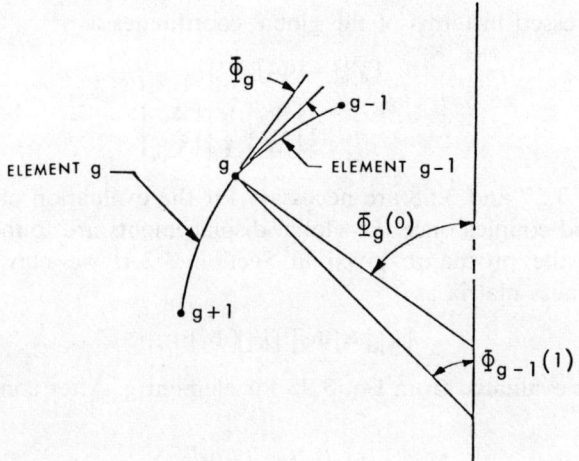

Fig. 3.3 Slope discontinuity at a node

in which
$$\Delta_{ga}^{i} = \Delta_{g}^{i}(0) = \Delta_{g-1}^{i}(1) = \Delta_{g-1,b}^{i} \quad (3.54a)$$
and
$$\Delta_{gb}^{i} = \Delta_{g}^{i}(1) \quad (3.54b)$$

However, the local nodal displacements at g for element g are
$$\Delta_{g}^{i}(0) = [\Phi_{1}]\Delta_{ga}^{i} \quad (3.55)$$
where
$$[\Phi_{1}] = \begin{bmatrix} \cos \bar{\phi}_{g} & 0 & -\sin \bar{\phi}_{g} & 0 & 0 \\ 0 & 1 & 0 & 0 & 0 \\ \sin \bar{\phi}_{g} & 0 & \cos \bar{\phi}_{g} & 0 & 0 \\ 0 & 0 & 0 & 1 & 0 \\ 0 & 0 & 0 & 0 & 1 \end{bmatrix} \quad (3.56a)$$
and
$$\bar{\phi}_{g} = \phi_{g}(0) - \phi_{g-1}(1) \quad (3.56b)$$

Equations 3.54 and 3.55 may be combined into a local displacement vector for element g as defined by Eq. 3.14:
$$\Delta_{g}^{i} = [\Phi_{2}]\bar{\Delta}_{g}^{i}$$
$$= \begin{bmatrix} \Phi_{1} & 0 \\ \hline 0 & \mathbf{I} \end{bmatrix} \begin{Bmatrix} \Delta_{ga}^{i} \\ \Delta_{gb}^{i} \end{Bmatrix} \quad (3.57)$$

Now, the vector of generalized displacements, as defined by Eq. 3.17,

Finite element analysis of shells of revolution

may be expressed in terms of the global coordinates as

$$\{\bar{Y}_g^i\} = [\Phi_3]\{\tilde{Y}_g^i\}$$
$$= \begin{bmatrix} \Phi_2 & 0 \\ 0 & \mathbf{I} \end{bmatrix} \begin{Bmatrix} \bar{\Delta}_g^i \\ \tilde{Y}_{ib}^i \end{Bmatrix} \quad (3.58)$$

Equations 3.57 and 3.58 are necessary for the evaluation of the stress resultants and couples once the global displacements are found.

Following the procedure given in Section 3.2.1, we may write the element stiffness matrix as

$$[k_g^i] = [\Phi_3]^T [k_i^i][\Phi_3] \quad (3.59)$$

where $[k_i^i]$ is evaluated from Eq. 3.25 for element g. After condensation, we have

$$[\bar{k}_g^i] = [\mathbf{k}_{gaa}^i - \mathbf{k}_{gab}^i \mathbf{k}_{gbb}^{i-1} \mathbf{k}_{gba}^i] \quad (3.60)$$

following Eqs 3.30–3.34. $[\bar{k}_g^i]$ is then used to assemble the global stiffness matrix at each node where a discontinuity occurs.

The nodal forces must also be expressed in terms of the common global coordinates. This is done by a transformation similar to Eq. 3.55,

$$\{\mathscr{F}_g^i\} = [\Phi_2]^T\{\mathscr{F}_i^i\} \quad (3.61)$$

in which $\{\mathscr{F}_i^i\}$ is evaluated from Eq. 3.35 for element g.

3.3.4 Branched shells

An extension to the concept of a compound shell, as discussed in the previous section, is illustrated on Fig. 3.4 where the branch *Brg* is shown

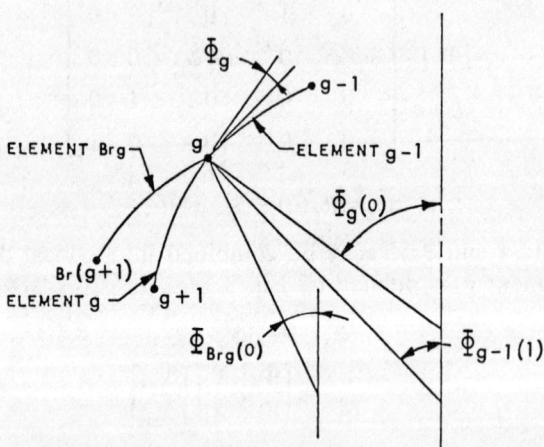

Fig. 3.4 Branched shell

emanating from node g. Commonly, the branch might be a straight or curved circumferential stiffener.

The analysis of the branched portion of the shell and the corresponding inclusion into the overall shell model may be conceptualized in the following steps:

(1) Modeling of the branch as a separate axisymmetric shell with the result being a global stiffness matrix $\bar{K}^j_{Brg}$ described in Section 3.2.3.1.
(2) Partitioning $\bar{K}^j_{Brg}$ and the corresponding displacement and load vectors into a form similar to Eq. 3.36 with the 'a' partition corresponding to node g and the 'b' partition to the rest of the branched shell.
(3) Following steps analogous to those employed in Section 3.3.3, resulting in a branch stiffness matrix and nodal force vector to be added to those of the main shell at node g.
(4) Renumbering the other nodes of the branch (in the 'b' partition) in accordance with the numbering system for the main shell.
(5) Solving the global problem.

In carrying out the last step, it should be recognized that the tightly banded form of the global equations will be altered substantially by the numbering of the nodes in the branch. Accordingly, it may be preferable to renumber the entire shell after the branches have been defined, with each branch numbered before proceeding to the next node. The implementation of a branched shell capability may represent a major modification to a basic axisymmetrical shell finite element program. However, many complex shells, otherwise either improperly or inadequately modeled by the basic assemblage of rings, can be elegantly addressed with this technique.[6]

3.3.5 Open-type elements

3.3.5.1 General

It is tempting and often expedient to treat shells which lack full axisymmetry but which still retain a circular planform using an axisymmetric model. A common example is the meridionally stiffened shell discussed in Section 3.3.2.3. A bolder extension of conventional shell theory is the application to circular frameworks which may be inserted at various levels in shell structures, as shown in Fig. 3.5. The upper openings may serve to admit light and air, while those at the lower boundary may also provide access. A prominent example is the system of inclined columns at the base of hyperbolic cooling towers which provide support when the shell is interrupted to admit the large volumes of air required by the system. If the members forming the framework are regular and closely spaced, it is

Finite element analysis of shells of revolution

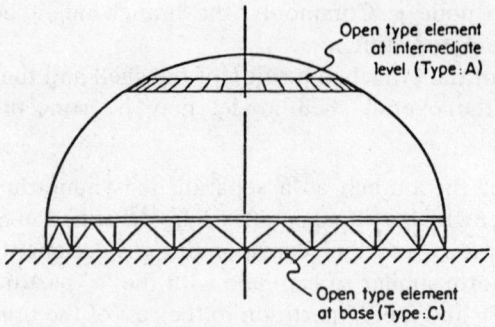

Fig. 3.5 Elliptical dome with column supports and roof openings

reasonable to smear the effect of each member over its tributary arc and then to treat the resulting system as an equivalent axisymmetric shell element. Such elements will be referred to as 'open-type elements', Fig. 3.6.

Beyond representing the basic stiffness of the columns, the modeling of open elements may include consideration of the connections between the columns and the shell, Fig. 3.7, and the inertia of the columns in the case of a dynamic analysis.[3]

3.3.5.2 Member displacement field

The following Hermitian polynomials in *local* Cartesian coordinates (x, y, z), which ensure displacement and slope continuity at the ends of

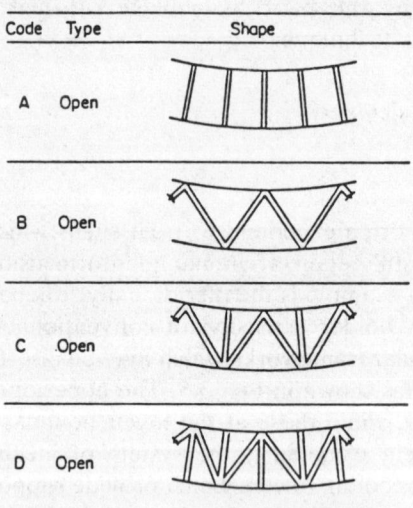

Fig. 3.6 Open-type elements

Static analysis

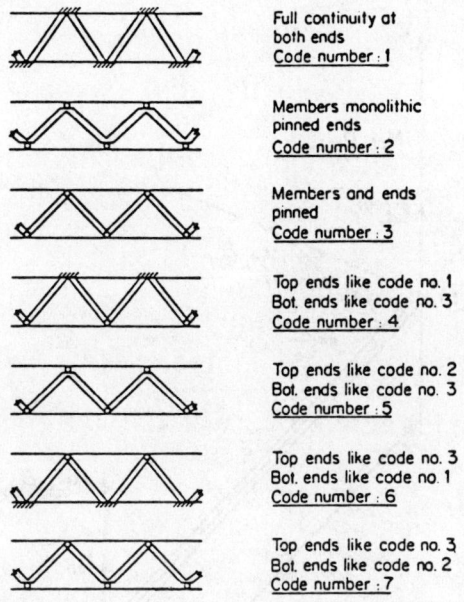

Fig. 3.7 End conditions for open-type elements

the members, are used:

$$\hat{u} = s\hat{u}_t + (1-s)\hat{u}_b$$
$$\hat{v} = (1-3s^2+2s^3)\hat{v}_t + (3s^2-2s^3)\hat{v}_b + s(s-1)^2 L_c \hat{B}_{zt} + s(s^2-s)L_c \hat{B}_{zb}$$
$$\hat{w} = (1-3s^2+2s^3)\hat{w}_t + (3s^2-2s^3)\hat{w}_b - s(s-1)^2 L_c \hat{B}_{yt} - s(s^2-s)L_c \hat{B}_{yb}$$
$$\hat{B}_x = s\hat{B}_{xt} + (1-s)\hat{B}_{xb} \tag{3.62}$$

in which $\hat{u}$, $\hat{v}$, $\hat{w}$ and $\hat{B}_x$ = the axial, lateral and normal displacements, and the rotation of the bar. The nodal variables $\hat{u}_t$, $\hat{v}_t$, $\hat{w}_t$, $\hat{B}_{yt}$, $\hat{B}_{zt}$, etc. are defined in Fig. 3.8 and the non-dimensional variable s is equal to x/L_c.

These displacement fields can be expressed as

in which
$$\{\hat{D}\} = [Z]\{\hat{\Delta}\} \tag{3.63a}$$

$$\{\hat{D}\} = \{\hat{u}\ \hat{v}\ \hat{w}\ \hat{B}_x\} \tag{3.63b}$$

and
$$\{\hat{\Delta}\} = \{\hat{u}_t \hat{v}_t \hat{w}_t \hat{B}_{xt} \hat{B}_{yt} \hat{B}_{zt} \hat{u}_b \hat{v}_b \hat{w}_b \hat{B}_{xb} \hat{B}_{yb} \hat{B}_{zb}\} \tag{3.63c}$$

$$[Z] = \text{a } (4 \times 12) \text{ connectivity matrix.} \tag{3.63d}$$

3.3.5.3 Member stiffness matrix

Following the approach of Sections 2.4.2 and 3.2.1, the linear strain–displacement relationship for a prismatic member will be of the form

$$\{\varepsilon\} = [\hat{B}]\{\hat{\Delta}\} \tag{3.64}$$

Finite element analysis of shells of revolution

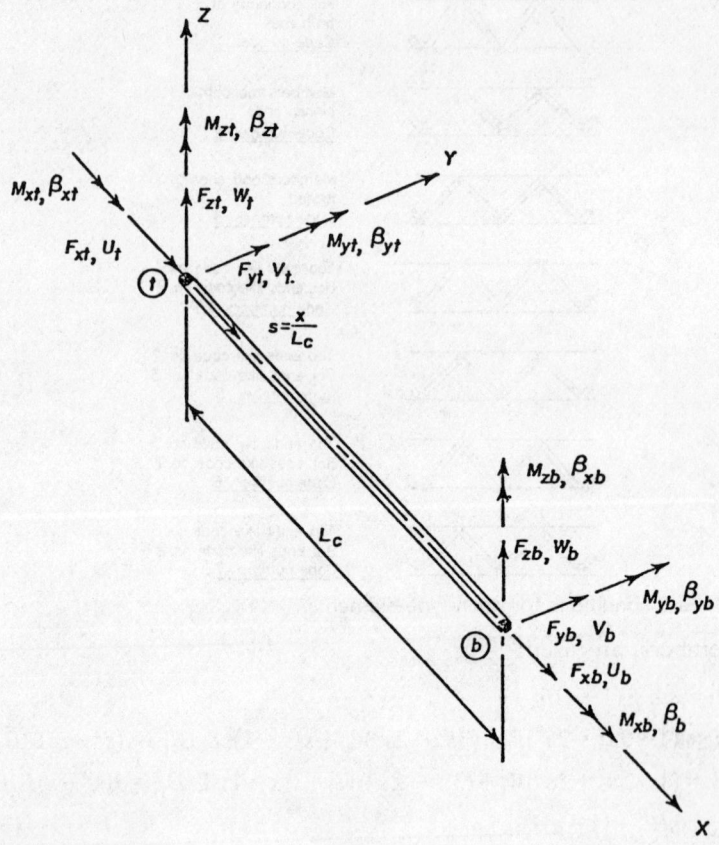

Fig. 3.8 Bar element

in which

$[\hat{B}]$ = the degree-of-freedom to strain transformation matrix

Then, the member stiffness matrix is expressed by

$$[\hat{k}] = L_c \int_0^1 [\hat{B}]^T [\hat{H}][\hat{B}] \, ds \qquad (3.65a)$$

in which the constitutive matrix, neglecting transverse shear and warping strains, is

$$[\hat{H}] = \begin{bmatrix} EA & 0 & 0 & 0 \\ 0 & EI_z & 0 & 0 \\ 0 & 0 & EI_y & 0 \\ 0 & 0 & 0 & GJ \end{bmatrix} \qquad (3.65b)$$

Static analysis

EA = the axial rigidity of the member
EI_z, EI_y = flexural rigidities of the member
GJ = the torsional rigidity of the member

Substituting $[\hat{B}]$ and $[\hat{H}]$ and carrying out the integrations, the resulting stiffness matrix with respect to the local coordinate system may be written as

$[\hat{k}] =$

$$\begin{bmatrix} \frac{EA}{L_c} & 0 & 0 & 0 & 0 & 0 & -\frac{EA}{L_c} & 0 & 0 & 0 & 0 & 0 \\ & \frac{12EI_z}{L_c^3}\delta_1 & 0 & 0 & 0 & \frac{6EI_z}{L_c^2}\delta_2 & 0 & -\frac{12EI_z}{L_c^3}\delta_1 & 0 & 0 & 0 & \frac{6EI_z}{L_c^2}\delta_3 \\ & & \frac{12EI_y}{L_c^3} & 0 & -\frac{6EI_y}{L_c^2} & 0 & 0 & 0 & -\frac{12EI_y}{L_c^3} & 0 & -\frac{6EI_y}{L_c^2} & 0 \\ & & & \frac{GJ}{L_c} & 0 & 0 & 0 & 0 & 0 & -\frac{GJ}{L_c} & 0 & 0 \\ & & & & \frac{4EI_y}{L_c} & 0 & 0 & 0 & \frac{6EI_y}{L_c^2} & 0 & \frac{2EI_y}{L_c} & 0 \\ & & & & & \frac{4EI_z}{L_c}\delta_4 & 0 & -\frac{6EI_z}{L_c^2}\delta_2 & 0 & 0 & 0 & \frac{2EI_z}{L_c}\delta_5 \\ \text{Symmetric} & & & & & & \frac{EA}{L_c} & 0 & 0 & 0 & 0 & 0 \\ & & & & & & & \frac{12EI_z}{L_c^3}\delta_1 & 0 & 0 & 0 & -\frac{6EI_z}{L_c^2}\delta_3 \\ & & & & & & & & \frac{12EI_y}{L_c^3} & 0 & \frac{6EI_y}{L_c^2} & 0 \\ & & & & & & & & & \frac{GJ}{L_c} & 0 & 0 \\ & & & & & & & & & & \frac{4EI_y}{L_c} & 0 \\ & & & & & & & & & & & \frac{4EI_z}{L_c}\delta_6 \end{bmatrix}$$

(3.66)

Equation 3.66 is identical to the local stiffness matrix for a member in a space frame, as given in Ref. 9, except for the coefficients δ_1 to δ_6 appearing in some of the elements of the stiffness matrix which have been introduced to account for the various members and conditions shown in Fig. 3.7. The values of these factors for different situations are tabulated in Table 3.1.

In Table 3.1, 'frame-pinned' means that the inclined members are monolithic but the connection with the rest of the structure is pinned about the z-axis, whereas 'members pinned' means that each member is pin-ended, as well.

Finite element analysis of shells of revolution

Table 3.1

End conditions		Code no.	Values of					
Top	Bottom		δ_1	δ_2	δ_3	δ_4	δ_5	δ_6
Fully continuous	Fully continuous	1	1	1	1	1	1	1
Frame-pinned	Frame-pinned	2	1	1	1	1	1	1
Members pinned	Members pinned	3	0	0	0	0	0	0
Fully continuous	Members pinned	4	$\frac{1}{4}$	$\frac{1}{2}$	0	$\frac{3}{4}$	0	0
Frame-pinned	Frame-pinned	5	$\frac{1}{4}$	$\frac{1}{2}$	0	$\frac{3}{4}$	0	0
Members pinned	Fully continuous	6	$\frac{1}{4}$	0	$\frac{1}{2}$	0	0	$\frac{3}{4}$
Members pinned	Frame-pinned	7	$\frac{1}{4}$	0	$\frac{1}{2}$	0	0	$\frac{3}{4}$

3.3.5.4 Coordinate transformations

Coordinate transformations may be required for two situations:

(i) In the case of inclined members, Fig. 3.9, it is necessary to transform the member matrices from the local coordinates to the curvilinear system (ϕ, θ, ζ), as shown on the figure. That is, the elements of $\{\Delta_i\}$ must be expressed in terms of the elements of $\{\hat{\Delta}_i\}$ which is defined in accordance with Eq. 3.14 as

$$\{\hat{\Delta}_i\} = \{u_t v_t w_t \beta_{\phi t} \beta_{\theta t} \beta_{\zeta t} u_b v_b w_b \beta_{\phi b} \beta_{\theta b} \beta_{\zeta b}\} \quad (3.67)$$

where t and b indicate the nodal circles at the top and bottom of the open element, respectively, as shown on Fig. 3.9.

(ii) In the event of a slope discontinuity, Fig. 3.10, it is necessary to further transform the preceding member matrices so as to correspond to the curvilinear coordinates of adjacent elements.

For inclined members, Case (i), the transformed stiffness matrix can be expressed by

$$[k_1] = [T_1]^T [\hat{k}][T_1] \quad (3.68a)$$

in which

$$[T_1] = \begin{bmatrix} \tau_1 & & & 0 \\ & \tau_1 & & \\ & & \tau_1 & \\ 0 & & & \tau_1 \end{bmatrix} \quad (3.68b)$$

$$[\tau_1] = \begin{bmatrix} c_1 & s_1 & 0 \\ -s_1 & c_1 & 0 \\ 0 & 0 & 1 \end{bmatrix} \quad (3.68c)$$

Static analysis

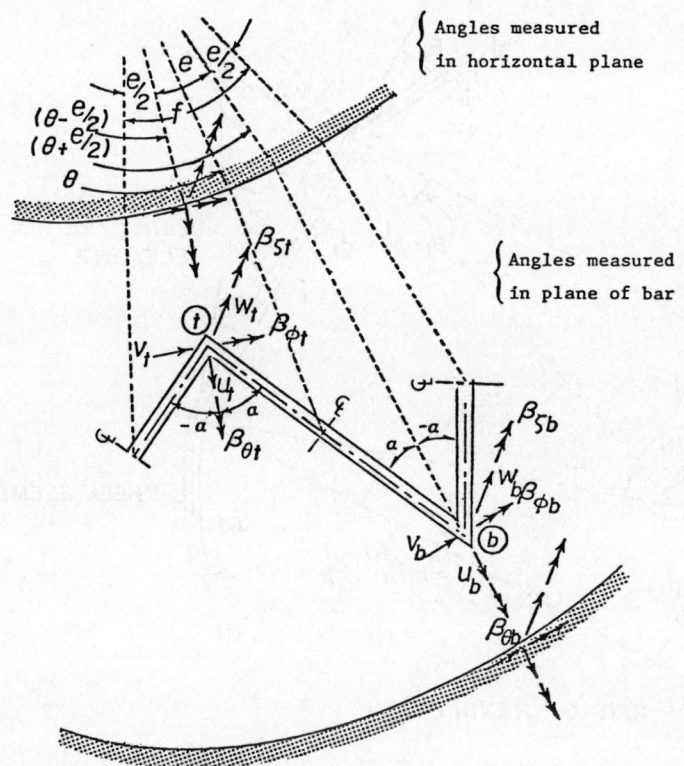

Fig. 3.9 Circumferential transformations

$$c_1 = \cos \alpha \tag{3.68d}$$
$$s_1 = \sin \alpha \tag{3.68e}$$

and α is the angle of inclination as shown on Fig. 3.9.

Then, if the middle surface of the open-type element introduces a slope discontinuity, Case (ii), the further transformation is

$$[\hat{k}_2] = [T_2]^T [\hat{k}_1][T_2] \tag{3.69a}$$

in which

$$[T_2] = \begin{bmatrix} \tau_{2t} & & & 0 \\ & \tau_{2t} & & \\ & & \tau_{2b} & \\ 0 & & & \tau_{2b} \end{bmatrix} \tag{3.96b}$$

$$[\tau_{2t}] = \begin{bmatrix} c_t & 0 & -s_t \\ 0 & 1 & 0 \\ s_t & 0 & c_t \end{bmatrix} \tag{3.69c}$$

47

Finite element analysis of shells of revolution

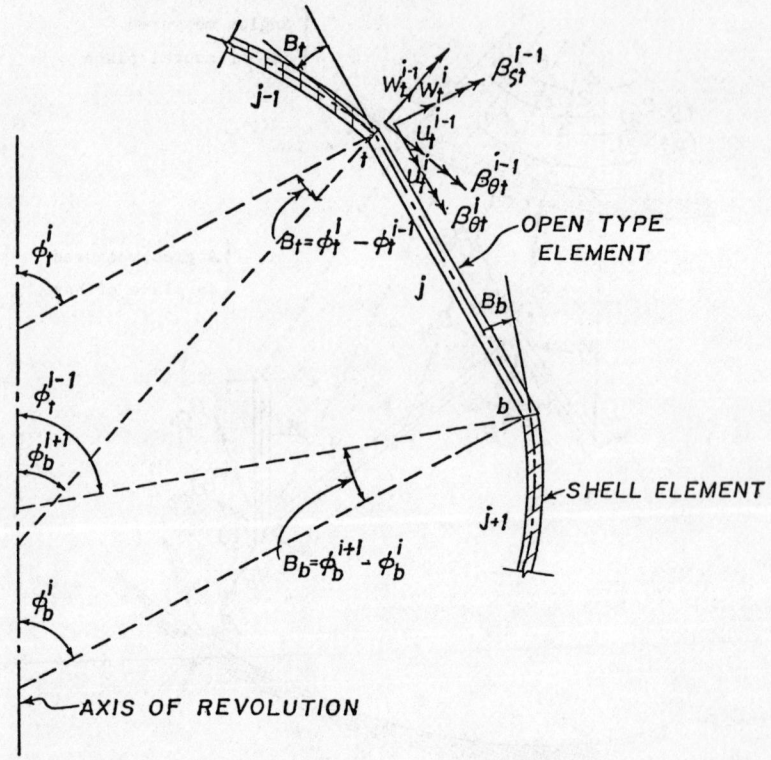

Fig. 3.10 Meridional transformations

$$c_t = \cos B_t \qquad (3.69d)$$

$$s_t = \sin B_t \qquad (3.69e)$$

$$\tau_{2b} = \tau_{2t} \text{ with } t \text{ replaced by } b \qquad (3.69f)$$

and B_t and B_b are the angles of departure from the meridian, as shown on Fig. 3.10.

The preceding transformations produce an (uncondensed) stiffness matrix (12×12) for an inclined member with a slope discontinuity at the nodes. If only transformation (i) is required $[\hat{k}_2] = [\hat{k}_1]$.

3.3.5.5 Reduced member stiffness matrix

It may be noted that the last two rows and columns of $[\hat{k}_2]$ correspond to rotations about the normal (z) axis at nodes t and b. In order to make these matrices consistent with those of the rotational shell elements (based, in general, on five degrees-of-freedom at each node), these two degrees-of-freedom (i.e., rotations about the normal) must be removed by condensation.

First the stiffness matrix is expressed in partitioned form as

$$[\hat{k}_2] = \begin{bmatrix} \hat{k}_{rr} & \vdots & \hat{k}_{rc} \\ (10\times 10) & \vdots & (10\times 2) \\ \hdashline \hat{k}_{cr} & \vdots & \hat{k}_{cc} \\ (2\times 10) & \vdots & (2\times 2) \end{bmatrix} \quad (3.70)$$

This matrix may now be condensed, identically to Eq. 3.34, into

$$[\hat{k}'] = [\hat{\mathbf{k}}_{rr} - \hat{\mathbf{k}}_{rc}\hat{\mathbf{k}}_{cc}^{-1}\hat{\mathbf{k}}_{cr}] \quad (3.71)$$

If the members of an open-type element are fully continuous with the adjacent shell elements, the contribution of the normal rotation degrees-of-freedom is neglected completely, in conformity with the basic assumptions of classical shell theories, and no condensation is necessary. In this case the reduced stiffness matrix is simply

$$[\hat{k}'] = \hat{\mathbf{k}}_{rr} \quad (3.72)$$

Equation 3.72 will also be applicable if the members have hinged connections with the adjacent shell elements, and if there is no slope discontinuity at the two nodes.

3.3.5.6 Effective stiffness matrix

Thus far, in the treatment of open-type elements, the circumferential variation of the displacement components has not been addressed. In order to take this into account, the displacement fields have to be expanded in Fourier series', as was done for the shell elements in Section 2.3.2 and particularly in Eq. 2.17.

Therefore, for a member with center located at θ measured in the horizontal plane, Fig. 3.9, the top and bottom end displacements of the member are written as

$$\{D_t\} = \sum_{j=0}^{\infty} \lceil \Theta_t^j \rfloor \{D_t^j\} \quad (3.73a)$$

$$\{D_b\} = \sum_{j=0}^{\infty} \lceil \Theta_b^j \rfloor \{D_b^j\} \quad (3.73b)$$

$$\lceil \Theta_t^j \rfloor = \lceil c_t^j s_t^j c_t^j c_t^j s_t^j \rfloor \quad (3.73c)$$

$$\lceil \Theta_b^j \rfloor = \lceil c_b^j s_b^j c_b^j c_b^j s_b^j \rfloor \quad (3.73d)$$

$$c_t^j = \cos j\left(\theta - \frac{e}{2}\right) \quad (3.73e)$$

$$s^i_t = \sin j\left(\theta - \frac{e}{2}\right) \qquad (3.73f)$$

$$c^i_b = \cos j\left(\theta + \frac{e}{2}\right) \qquad (3.73g)$$

$$s^i_b = \sin j\left(\theta + \frac{e}{2}\right) \qquad (3.73h)$$

where e is the horizontal angle subtended by the member at the axis of revolution, as shown in Fig. 3.9. For members oriented along the shell meridian, $e = 0$. The effective stiffness matrix can now be derived by[3]

(1) smearing the properties of one or more members over their tributary arc; and
(2) summing the effect of all members around the circumference by integrating with respect to θ. In the case of inclined members, it is more convenient to consider a sub-assemblage consisting of one full member and the adjacent half-members, which subtends an angle f equal to $2e$ at the axis of revolution, as shown in Fig. 3.9. In the cases of members oriented along the meridian, f will be equal to the angular spacing of the members.

In Fig. 3.9, if the angle α between the member t–b and the meridian is taken as positive, then the value for the adjacent members will be negative and, accordingly, the $\sin \alpha$ (or s_1) terms appearing in the stiffness matrix for these members will have opposite signs. Therefore, in the case of inclined members, the elements of the effective stiffness matrix $[\tilde{k}]$ for the location θ, valid over the circumferential angle $f = 4/N$ where N is the total number of members, will be as follows:

$$\begin{bmatrix} \tilde{k}_{i,i} = 2\hat{k}'_{i,i} & (i = 1, 10) & & & \\ \tilde{k}_{1,2} = 0; & \tilde{k}_{1,3} = 2\hat{k}'_{1,3}; & \tilde{k}_{1,4} = 2\hat{k}'_{1,4}; & \tilde{k}_{1,5} = 0 & \\ \tilde{k}_{2,3} = 0; & \tilde{k}_{2,4} = 0; & \tilde{k}_{2,5} = 2\hat{k}'_{2,5}; & \tilde{k}_{3,4} = 2\hat{k}'_{3,4} & \\ \tilde{k}_{3,5} = 0; & \tilde{k}_{4,5} = 0; & \tilde{k}_{6,7} = 0; & \tilde{k}_{6,8} = 2\hat{k}'_{6,8} & \\ \tilde{k}_{6,9} = 2\hat{k}'_{6,9}; & \tilde{k}_{6,10} = 0; & \tilde{k}_{7,8} = 0; & \tilde{k}_{7,9} = 0 & \\ \tilde{k}_{7,10} = 2\hat{k}'_{7,10}; & \tilde{k}_{8,9} = 2\hat{k}'_{8,9}; & \tilde{k}_{8,10} = 0; & \tilde{k}_{9,10} = 0 & \end{bmatrix} \qquad (3.74)$$

The rest of the terms will be unchanged. In the case of meridionally oriented members, no modification of the matrices is necessary. Now, after dividing the elements of these matrices by f, the stiffness matrix corresponding to a unit circumferential angle is obtained.

For each harmonic, the final form of the stiffness matrix for an open-type element is obtained by using Eqs 3.73 and 3.74 and integrating over the circumference as[3]

$$[\bar{k}_{op}^i] = \frac{1}{f}\int_{-\pi}^{\pi} [\tilde{\Theta}][\tilde{k}][\tilde{\Theta}]\,d\theta \qquad (3.75)$$

in which

$$[\tilde{\Theta}] = \lceil \Theta_t^i \Theta_b^i \rfloor \qquad (3.76)$$

where the deformations are taken as symmetrical about the $\theta = 0$ axis.

Noting that

$$\int_{-\pi}^{\pi} (c_t^i)^2\,d\theta = \int_{-\pi}^{\pi} (s_t^i)^2\,d\theta = \pi \qquad (3.77a)$$

$$\int_{-\pi}^{\pi} s_t^i c_t^i\,d\theta = \int_{-\pi}^{\pi} s_b^i c_b^i\,d\theta = 0 \qquad (3.77b)$$

$$\int_{-\pi}^{\pi} s_b^i s_t^i\,d\theta = \int_{-\pi}^{\pi} c_b^i c_t^i\,d\theta = \pi \cos je \qquad (3.77c)$$

$$\int_{-\pi}^{\pi} s_b^i c_t^i\,d\theta = \int_{-\pi}^{\pi} s_t^i c_b^i\,d\theta = \pi \sin je \qquad (3.77d)$$

the effect of integration around the circumference on $[\tilde{k}]$ in Eq. 3.75 may be summarized as follows:[3]

1. All diagonal terms are multiplied by $\lambda \pi$.
2. The terms $\tilde{k}_{1,3}$, $\tilde{k}_{1,4}$, $\tilde{k}_{2,5}$, $\tilde{k}_{3,4}$, $\tilde{k}_{6,8}$, $\tilde{k}_{7,10}$ and $\tilde{k}_{8,9}$ are multiplied by $\lambda \pi$.
3. The terms $\tilde{k}_{1,6}$, $\tilde{k}_{1,8}$, $\tilde{k}_{1,9}$, $\tilde{k}_{2,7}$, $\tilde{k}_{2,10}$, $\tilde{k}_{3,6}$, $\tilde{k}_{3,8}$, $\tilde{k}_{3,9}$, $\tilde{k}_{4,6}$, $\tilde{k}_{4,8}$, $\tilde{k}_{4,9}$, $\tilde{k}_{5,7}$ and $\tilde{k}_{5,10}$ are multiplied by $\lambda \pi \cos je$.
4. The terms $\tilde{k}_{1,7}$, $\tilde{k}_{1,10}$, $\tilde{k}_{3,7}$, $\tilde{k}_{3,10}$, $\tilde{k}_{4,7}$ and $\tilde{k}_{4,10}$ are multiplied by $\lambda \pi \sin je$.
5. The terms $\tilde{k}_{2,6}$, $\tilde{k}_{2,8}$, $\tilde{k}_{2,9}$, $\tilde{k}_{5,6}$, $\tilde{k}_{5,8}$ and $\tilde{k}_{5,9}$ are multiplied by $-\lambda \pi \sin je$.

For $j = 0$, $\lambda = 2$ and for $j \geq 1$, $\lambda = 1$.

The matrix $[\bar{k}_{op}^i]$ may then be used interchangeably with the shell element stiffness matrix $[\bar{k}_i^j]$ as given by Eq. 3.34.

3.3.5.7 Consistent load vectors

Along with the development of a specialized element stiffness matrix, it is appropriate to properly represent the loading for open-type elements. Load vectors have been developed for self-weight, thermal gradient and normal pressure loading.[3] As they are of specialized interest, the expressions are not repeated here; however, for many applications, a sufficiently accurate approximation may be obtained by calculating tributary reactions on the open elements and applying these reactions as ring

Finite element analysis of shells of revolution

loads at the junction of the open and shell elements. Such ring loads are of the form given by Eq. 2.27.

For the cases noted, it is probably sufficient to consider the reactions to be uniformly applied around the circumference and thus restrict considerations to the $j = 0$ harmonic.

3.3.6 Local effects of columns on shell

The open-type elements derived in the previous section enable the effect of the discrete column supports on the overall response of the shell to be included. However, the necessity of smearing the effects over the tributary arc suppresses the local amplification of stresses in the immediate vicinity of the column-shell intersection. This may be rectified by a correction whereby the effects of the discrete boundary reactions are represented by a superposition procedure, Fig. 3.11, originally conceived by Aas-Jakobsen for symmetrically loaded spherical shells;[10] generalized to nonsymmetrically loaded shells by Gould and Lee;[11] and adapted to finite element analysis by El-Shafee and Gould.[12]

The most prominent stress resultant concentrated by the column supports, N_ϕ, is selected for purposes of illustration, but the same treatment is used for all interrupted stress resultants and couples. The *discrete* reactions, $\tilde{N}_\phi$, Fig. 3.11(a), are assumed to be uniformly distributed over the circumferential width of the column and equal in magnitude to the tributary *continuous* reaction N_ϕ, Fig. 3.11(b). The effects of the discrete reactions, (a), are then evaluated as the superposition of the results of an analysis for the applied surface loading with the boundary *continuously* supported, (b); and an analysis with only the edge loading applied, (c). The latter case is self-equilibrated, and consists of the *negative* of case (b) *between* columns and the *difference* of cases (a) and (b) *within* the column width.

The first step is to evaluate the column reaction intensities from the case (b) solution for a column with center line at $\theta = \theta_\lambda$:

$$\tilde{N}_{\lambda i} = \tilde{N}_i(\theta_\lambda) = \frac{1}{2\beta R_b} \int_{\theta_\lambda - (\pi/c)}^{\theta_\lambda + (\pi/c)} N_i(\phi_s, \theta) R_b \, d\theta \qquad (3.78)$$

in which:

c = total number of equally spaced column support points
λ = column number $(0, 1, \ldots, c-1)$

$$\theta_\lambda = \frac{2\pi\lambda}{c}$$

2β = angle subtended by column measured in the horizontal plane
R_b = base radius

Static analysis

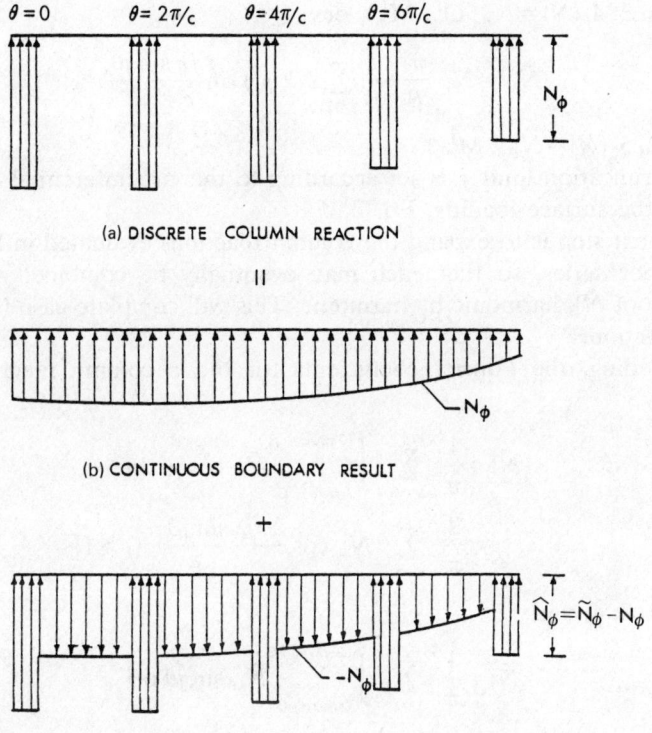

Fig. 3.11 Discrete column analysis

and

$$N_i = N_\phi, N_{\theta\phi}, Q_\phi, M_\phi, M_{\theta\phi} \qquad (3.79)$$
$$= N_1, N_2, N_3, N_4, N_5$$
= Continuous boundary stress resultants and couples at the lower boundary of shell ϕ_s, as defined in Eq. 2.23.

The units of $\tilde{N}_{\lambda i}$ are the same as those of the corresponding stress resultants N_i, force per length.

In order to evaluate Eq. 3.78, the continuous boundary stress resultants N_i must be known. These are obtained from the Case (b) solution, and it is assumed that they are available in the form of Eq. 2.23 with Fourier coefficients N_i^j given by Eq. 2.24.

Substituting $N_i^j \cos j\theta_\lambda$ for element i into Eq. 3.78 and integrating gives

$$\tilde{N}_{\lambda i} = N_i^0 \frac{\pi}{c\beta} + \frac{1}{\beta} \sum_{j=1}^{\bar{j}} N_i^j(\phi_s) \sin\frac{j\pi}{c} \frac{\cos j\theta_\lambda}{j} \qquad (3.80a)$$

Finite element analysis of shells of revolution

for $i = 1, 3, 4$ $(N_i^j = N_\phi^j, Q_\phi^j, M_\phi^j)$, or

$$\tilde{N}_{\lambda i} = N_i^1 \frac{\pi}{c\beta} + \frac{1}{\beta} \sum_{j=2}^{\bar{j}} N_i^j(\phi_s) \sin \frac{j\pi}{c} \frac{\sin j\theta_\lambda}{j} \tag{3.80b}$$

for $i = 2, 5$ $(N_i^j = N_{\theta\phi}^j, M_{\theta\phi}^j)$.

The truncation limit $\bar{j}$ is set according to the circumferential distribution of the surface loading, Eq. 2.25.

The next step is to expand the column reactions evaluated in Eq. 3.80 in Fourier series, so that each may eventually be combined with the negative of N_i^j, harmonic by harmonic. This will complete case (c) of the superposition.

Proceeding, the Fourier coefficients for the c column reactions are given by

$$\tilde{N}_i^j = \frac{1}{\pi} \sum_{\lambda=0}^{\lambda=c-1} \int_{(2\pi\lambda/c)-\beta}^{(2\pi\lambda/c)+\beta} \tilde{N}_{\lambda i} \cos j\theta \, d\theta$$

$$= \frac{2}{\pi} \sum_{\lambda=0}^{\lambda=c-1} \tilde{N}_{\lambda i} \cos \frac{2\pi j \lambda}{c} \frac{\sin j\beta}{j} \quad (j > 1) \tag{3.81a}$$

for $i = 1, 3, 4$, and

$$\tilde{N}_i^j = \frac{1}{\pi} \sum_{\lambda=0}^{\lambda=c-1} \int_{(2\pi\lambda/c)-\beta}^{(2\pi\lambda/c)+\beta} \tilde{N}_{\lambda i} \sin j\theta \, d\theta$$

$$= \frac{2}{\pi} \sum_{\lambda=0}^{\lambda=c-1} \tilde{N}_{\lambda i} \sin \frac{2\pi j \lambda}{c} \frac{\sin j\beta}{j} \quad (j > 1) \tag{3.81b}$$

for $i = 2, 5$. Then, case (c) of the superposition is given by the Fourier coefficients:

$$\hat{N}_i^j = \tilde{N}_i^j - N_i^j \qquad (j \le \bar{j}) \tag{3.82a}$$

and

$$\hat{N}_i^j = \tilde{N}_i^j \qquad (\bar{\bar{j}} \ge j > \bar{j}) \tag{3.82b}$$

The truncation limit for the expansion of the column reactions $\bar{\bar{j}}$ should, in general, be much larger than $\bar{j}$, which was used for the continuous boundary reactions in Eqs 3.80. For symmetrical applied loading, it may be shown that only *integer* multiples of the total number of columns, $j = n \times c$, give non-zero values of $\tilde{N}_i^j$, while for non-symmetric loading, it may be necessary to consider several hundred harmonics to adequately describe the reaction. From a practical standpoint, it is usually advisable to reduce the total number of $\tilde{N}_i^j$ before performing the subsequent stress analysis. One possibility is to retain only those values which are a reasonable fraction (say 5–10%) of the maximum value of a particular $\tilde{N}_i^j$.

The final step is to introduce the selected Fourier coefficients represent-

Static analysis

ing case (c) as circumferential line loads at $\phi = \phi_s$ and to perform a finite element analysis harmonic by harmonic. Presumably, this may be accomplished with the same computer code used for case (b) to determine N_i^j. In describing this model, however, it may be efficient to alter the finite element grid somewhat from that used in case (b) since the case (c) stresses and displacements tend to attenuate very rapidly from the lower boundary. As case (c) is self-equilibrating, no kinematic constraints are required.

It is also possible to use this technique to approximately correct the results of a dynamic analysis for the local column effects. This will be demonstrated in the case study in Chapter 7.

3.4 Mixed formulation

We refer to the functional I defined by Eq. 2.61 and the approximations over the element, Eqs 3.6–3.13. Then, following steps analogous to those employed in Section 3.2.1 to derive the equilibrium equations and in Section 3.2.2 to reduce these equations, the condensed mixed equations for element i and harmonic j may be derived in the form[13]

$$[\bar{m}_i^j]\{\bar{Y}_{ia}^j\} = \{\bar{\mathscr{F}}_i^j\} \tag{3.83}$$

in which

$[\bar{m}_i^j]$ = the element mixed matrix

$$\{\bar{Y}_{ia}^j\} = \{u_i^j(0)v_i^j(0)w_i^j(0)m_{\phi i}^j(0)m_{\theta i}^j(0)m_{\theta\phi i}^j(0)$$
$$u_i^j(1)v_i^j(1)w_i^j(1)m_{\phi i}^j(1)m_{\theta i}^j(1)m_{\theta\phi i}^j(1)\} \tag{3.84}$$

$\{\bar{\mathscr{F}}_i^j\}$ = the vector giving the contribution of the loads acting on element i to the equivalent nodal forces at nodes i and $i+1$.

The solution format is identical to that used for the displacement formulation; however, the resultant system of equations is *not* positive–definite so that many popular equation solvers cannot be used. However, standard Gauss elimination methods are applicable to such equations.

Apparent advantages of the mixed solution are that continuous stress couples result at the nodes and that low order interpolations (linear and quadratic) have been shown to give excellent results. This must be balanced against the increase in nodal variables from 6 to 5 (Eqs 3.84 versus 3.14). Comparative examples between the two methods will be presented subsequently.

3.5 Convergence and discretization criteria

3.5.1 Convergence

The topic of convergence of finite element solutions (and numerical methods in general) is widely treated in the mathematical and engineering literature. Currently, there is considerable discussion as to whether the conventional approach of achieving convergence by refining the discretization mesh while holding the element order constant (h-convergence) is more efficient than an approach whereby the order of approximation within the element is refined while holding the mesh constant (p-convergence).[14] The resolution of this question is beyond the goals of this book, but the element derived herein permits p as well as h convergence to be exploited as the order of approximation is variable, e.g. Eq. 3.2. In practice, for the element types considered herein, it has been expedient to utilize p-convergence up to a preset limit, chosen as $n = 6$, and then turn to traditional h convergence.

A further aspect is the precise measure of convergence when analytical or other independent solutions are not available. A *single* measure of the quality of one solution against another is the value of the functional being minimized, the smaller value indicating a better solution in a global sense. Comparison of key results, e.g., displacements and stress resultants, between two solutions is helpful, with a close replication of the former values by a later refinement signaling convergence. An additional check, which is well suited for rotational shell elements, is the interelement continuity of stress resultants and couples. Since these values are independently computed for the adjacent elements and involve differentiation of the displacement functions, they are quite sensitive and the achievement of close interelement continuity has been a reliable indicator of convergence. This form of check may not be available from all programs, however, since results are commonly sampled only at the Gauss points which are in the interior away from the boundaries.

3.5.2 Discretization criteria

It is expedient to improve the mesh arrangement in a systematic manner. One technique that has been employed is to consider available analytical solutions for shells subjected to edge loading and to study the attenuation of the stress field away from the loaded edge.[15] In Ref. 15, it was suggested that the decay length for a given shell in the vicinity of a local intensification be approximated by first computing

$$X_c = \frac{\pi\sqrt{(Rh)}}{\sqrt[4]{[3(1-\mu^2)]}} \qquad \phi \geq 40° \qquad (3.85a)$$

Static analysis

or

$$X_c = \frac{\pi^2 h \cos \phi}{\sqrt{3[(1-\mu^2)]}} \qquad \phi < 40° \qquad (3.85b)$$

Equation 3.85(a) is derived from an edge-loaded cylindrical shell, while Eq. 3.85(b) is based on a similarly loaded conical shell. Then, the length

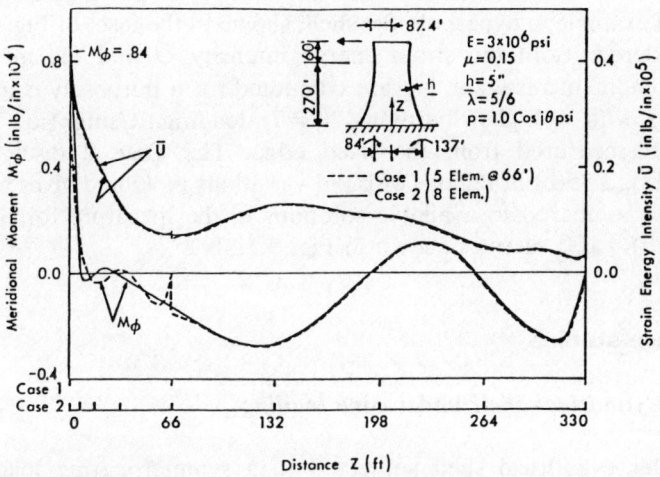

Fig. 3.12(a) Distribution of $\bar{U}$ and M_ϕ in a hyperboloidal shell under seventh harmonic normal loading ($j=7$)

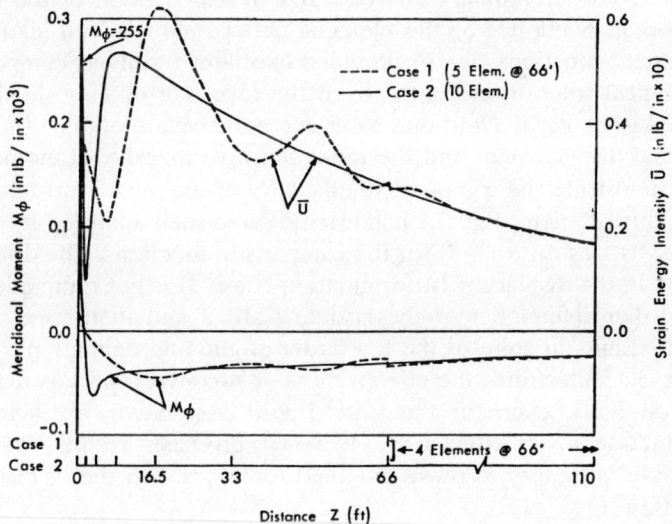

Fig. 3.12(b) Distribution of $\bar{U}$ and M_ϕ in a hyperboloidal shell under one-hundredth harmonic normal loading ($j=100$)

of the smallest element to be employed within X_c is

$$L_i = X_c/2 \tag{3.86a}$$

for $j = 0$ and

$$L_i = X_c/2^p \tag{3.86b}$$

where $p \simeq \sqrt[3]{j}$.[15] Both formulae are for the order of polynomial $n = 6$.

As an example, a hyperboloidal shell, shown in the inset of Fig. 3.12(a), is considered. Both the strain energy intensity $\bar{U}$ and the meridional bending moment resultant M_ϕ are computed for a purposely coarse grid (Case 1) with a higher harmonic ($j = 7$) loading. Using Eq. 3.85(a), $X_c = 15$ ft measured from the fixed edge. The Case 2 discretization utilizes Eq. 3.85(b) in regions of rapid variations of $\bar{U}$ and gives accurate results as compared to available solutions in the literature for both the $j = 7$ and $j = 100$ cases as shown in Fig. 3.12(b).[15]

3.6 Case studies

3.6.1 Cylindrical shell under edge loading

A circular cylindrical shell subjected to a symmetric ring loading, as shown on the inset of Fig. 3.13, is an informative example to focus on discretization patterns since doubly curved geometry is not an issue. Based on some preliminary analyses and the application of the criteria developed in Section 3.5, the element pattern shown with sixth-order displacement functions ($n = 6$) provided excellent results as compared to the analytical solution evaluated from the formulae in Timoshenko and Woinowski-Kreiger.[16] Note that close comparisons are obtained for both the normal displacement and the more sensitive meridional moment.

To demonstrate the comparative efficiency of the mixed formulation in treating this problem, Fig. 3.14 shows the same shell analyzed using only quadratic expansions ($n = 2$) for the comparison functions. The discretization used in the displacement formulation (Case 1) is not quite adequate, but a further refinement of the grid to Case 2 and then Case 3 gives excellent results in spite of the low order of the approximation.

Figure 3.15 illustrates the effectiveness of p-convergence, as described in Section 3.5.1, even for the Case 1 grid. As shown, an increase in approximation order from $n = 2$ to $n = 4$ produces results which are comparable in quality to those obtained for $n = 6$ with the displacement formulation, Fig. 3.13.

The jump discontinuity for the transverse shearing force Q_ϕ using the Case 1 discretization and quadratic approximation is attributed to the

Static analysis

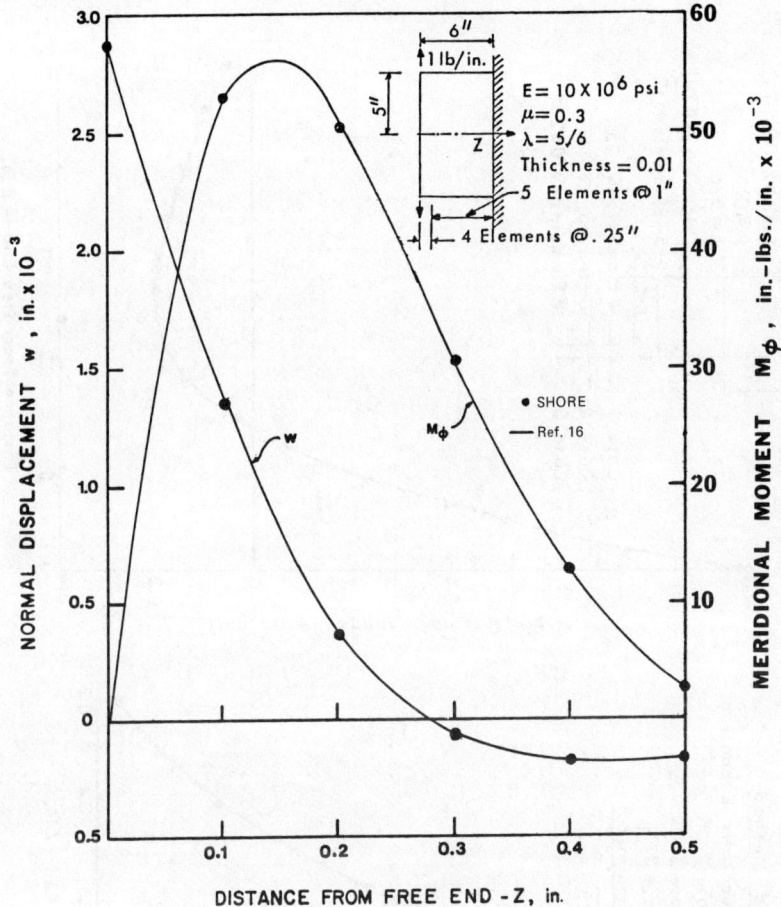

Fig. 3.13 w and M_ϕ versus Z—edge-loaded cylindrical shell

sensitivity of this term, which is evaluated by differentiation from the comparison functions. The discontinuity disappears with refinement of the grid.

3.6.2 Cylindrical shell under hydrostatic loading

The purposes of this example are to show that distributed loadings which vary in the axial direction may be accurately represented, and to numerically measure the influence of transverse shearing strains. In Fig. 3.16, SHORE solutions for transverse shear Q_ϕ and meridional moment M_ϕ are plotted for shear stress shape factors λ equal to 5/6 and 50. The

59

Finite element analysis of shells of revolution

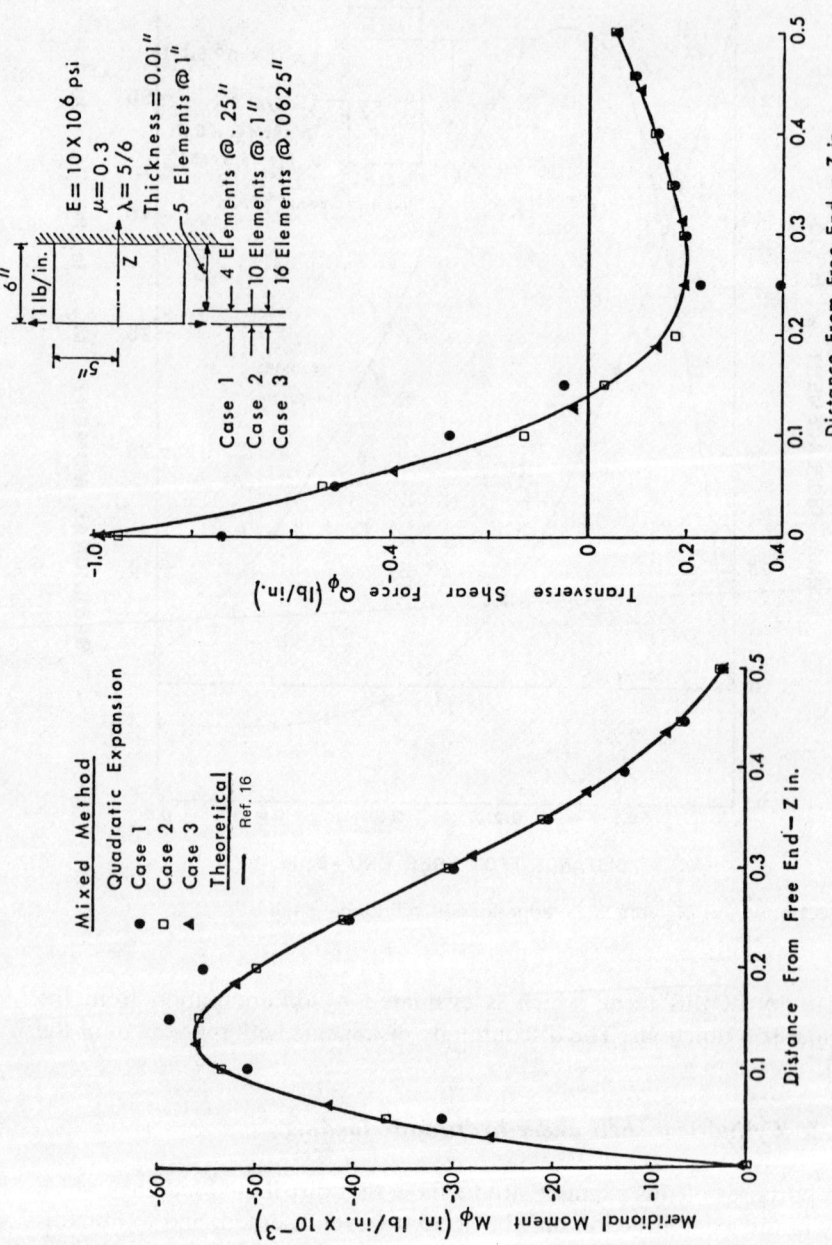

Fig. 3.14 Edge-loaded cylindrical shell

Static analysis

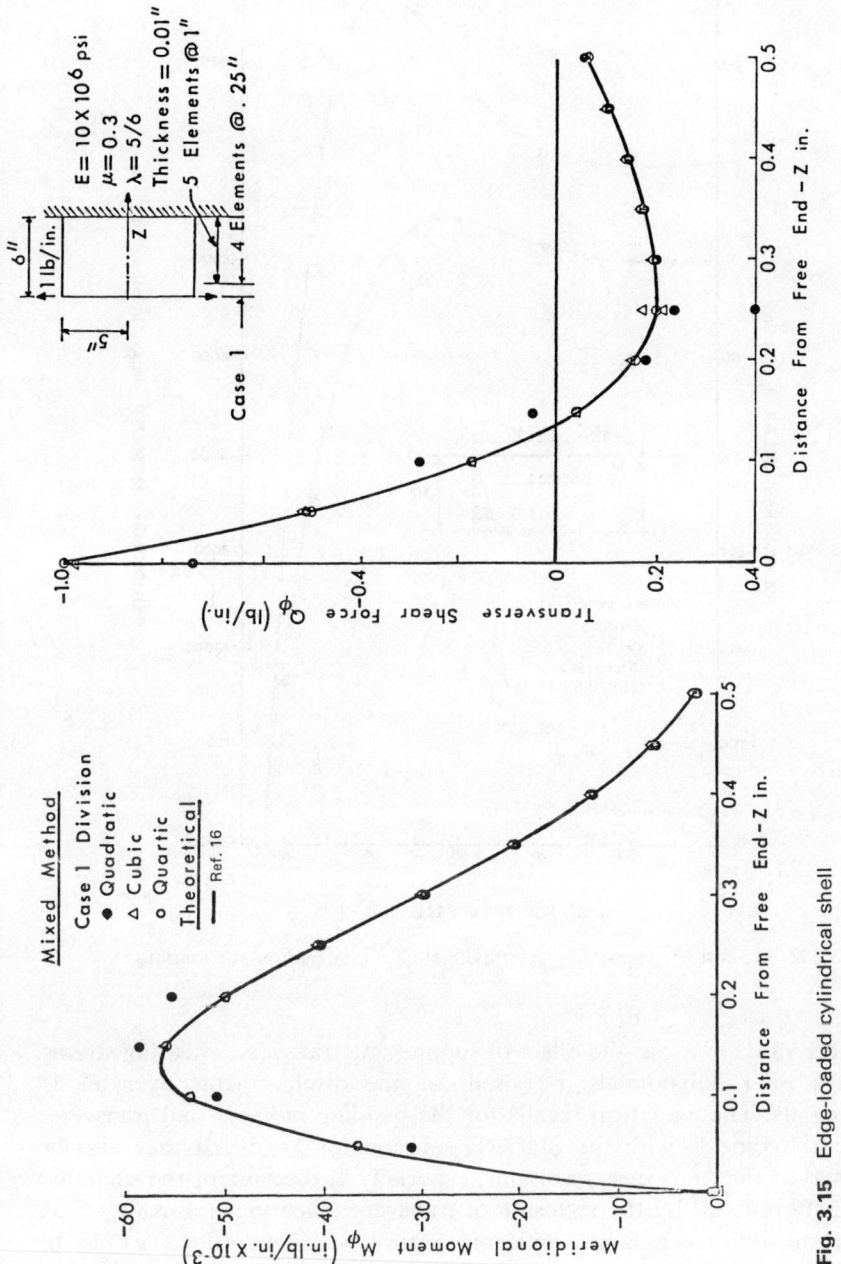

Fig. 3.15 Edge-loaded cylindrical shell

Finite element analysis of shells of revolution

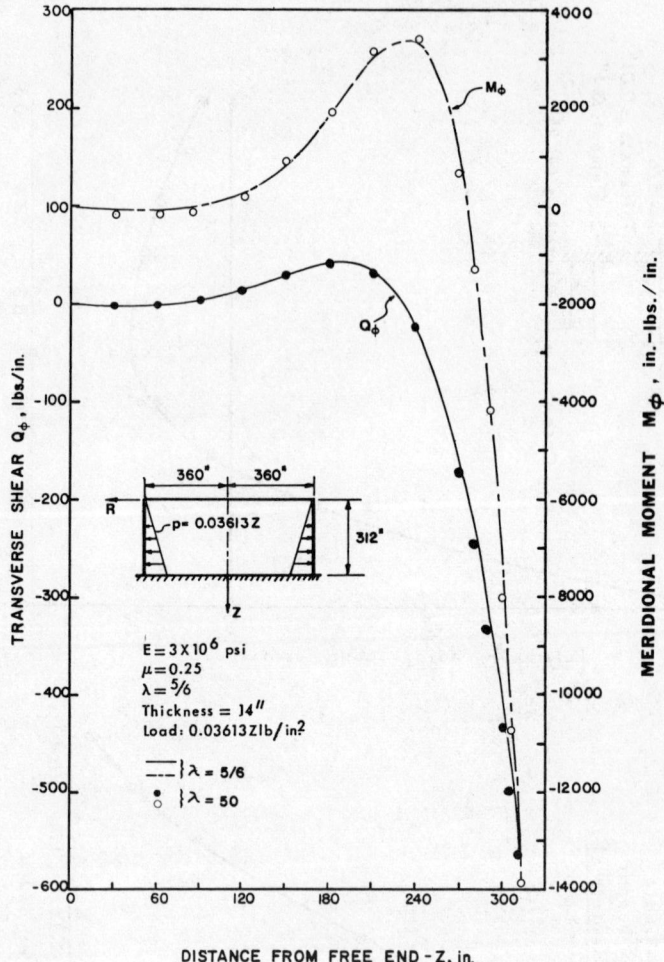

Fig. 3.16 Q_ϕ and M_ϕ versus Z—cylindrical shell under hydrostatic loading

larger value of λ has the effect of suppressing transverse shearing strains. Sixth-order polynomials are used for the displacements over all 14 elements. The analytical results for the bending moment and transverse shear[16] coincide with the SHORE solution for $\lambda = 50$. It may also be observed that the stress resultants, especially at the base of the shell, are slightly reduced by the inclusion of transverse shearing strains ($\lambda = 5/6$), but the difference is insignificant. Transverse shear effects would be expected to be more prominent for cases with concentrated loading and/or geometrical discontinuities.

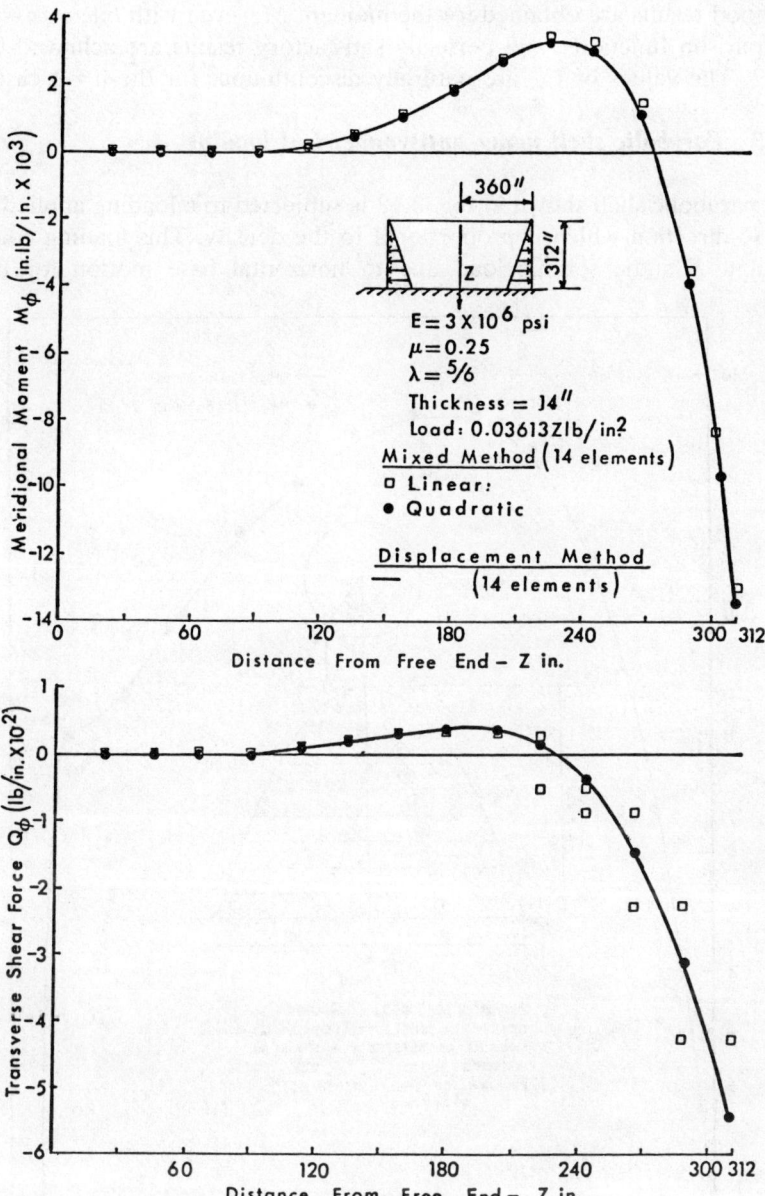

Fig. 3.17 Cylindrical shell under hydrostatic loading

Finite element analysis of shells of revolution

In Fig. 3.17 the solution is based on the mixed formulation. Remarkably good results are obtained for the moment M_ϕ, even with *linear* ($n = 1$) comparison functions, and perfectly satisfactory results are achieved for $n = 2$. The values of Q_ϕ are naturally discontinuous for the $n = 1$ case.

3.6.3 Parabolic shell under antisymmetrical loading

The parabolic shell shown in Fig. 3.18 is subjected to a loading applied in the R direction which is proportional to the density. This loading could simulate a static seismic load due to horizontal base motion for the

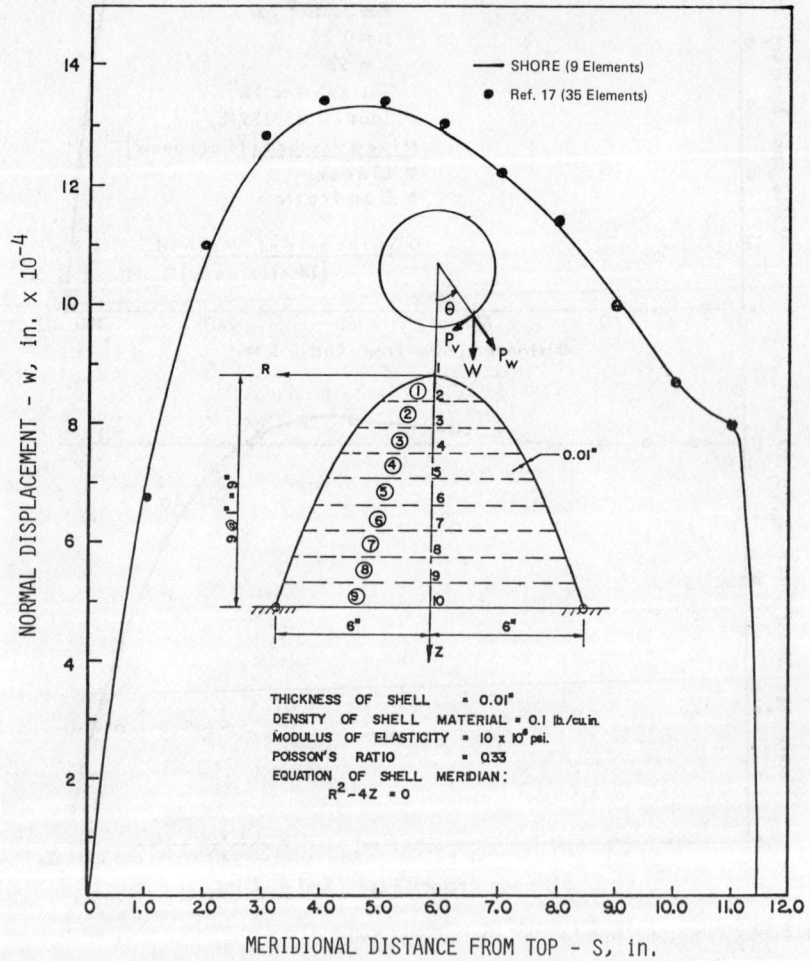

Fig. 3.18 Normal displacement w versus S—parabolic shell under antisymmetrical radial loading

Static analysis

Z-axis oriented vertically, or the self-weight of the shell if the shell were cantilevered with the Z-axis horizontal.

This loading condition corresponds to $f_w = W \cos \theta$ and $f_v = -W \sin \theta$ where $W =$ the weight of the shell (or some proportion thereof), and thus involves only the $j = 1$ harmonic. In the case illustrated $W = 0.1 \times 0.01 = 0.001 \text{ lb/in}^2$. Also, the first element is necessarily a cap element. Only

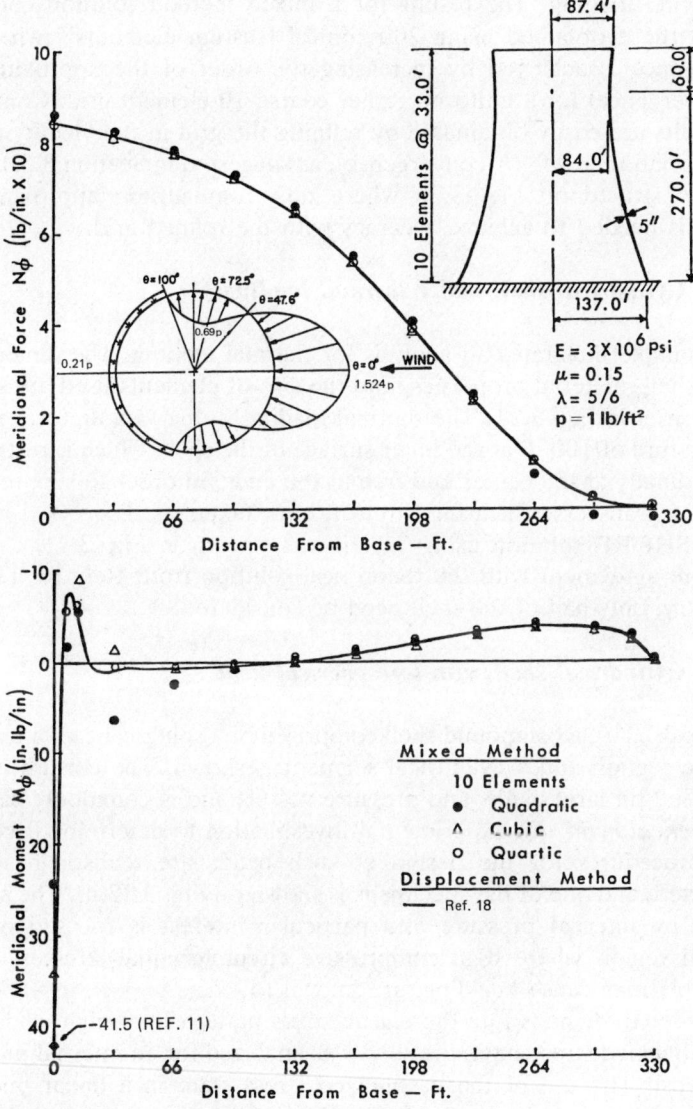

Fig. 3.19 Hyperboloidal shell under static wind loading.

Finite element analysis of shells of revolution

limited results are available for verification;[17] however, the displacement comparison seems good.

3.6.4 Hyperboloidal shell under static wind loading

The analysis of a non-symmetrically loaded shell is summarized in Fig. 3.19. Nine harmonics ($j = 0$ to 8) are required to adequately represent the static wind loading. The results for a mixed method solution compare well to those obtained using 200 conical frustum elements[18] when the convergence is achieved by increasing the order of the approximation (p-convergence) for a uniform, rather coarse 10 element grid. Comparable results are easily obtainable by refining the grid in the vicinity of the rapid variation in M_ϕ (h-convergence), as suggested in Section 3.5.1. This is demonstrated on Fig. 3.20 where only a quadratic approximation ($n = 2$) is needed to achieve accuracy with the refined grid.

3.6.5 Cylindrical shell under thermal loading

This example illustrates an analysis for thermal loading. The dimensions of the shell, material properties, and the size of elements used are shown on the inset in Fig. 3.21. The thermal loading is due to a uniform rise in temperature of 100 °F at the inner surface of the shell, which is restrained longitudinally at the center and free at the ends. In order to suppress the effect of transverse shearing strains, λ is taken as 100.

The SHORE solution using 7 elements, shown in Fig. 3.21, exhibits complete agreement with the theoretical solution from Ref. 16. Due to symmetry, only half of the shell need be considered.

3.6.6 Cylindrical shell with torospherical head

On Fig. 3.22(a), a compound shell comprised of a spherical cap, a toroidal knuckle section and a cylindrical segment is shown. The closure is of a form used on large tanks and pressure vessels and is commonly called a *torospherical head*. An experimental investigation to determine if current code procedures for the design of such heads are realistic has been conducted, and one of the specimens is shown on Fig. 3.22(b). The shell is loaded by internal pressure and particular interest is focused on the toroidal region where high compressive circumferential stresses, which ostensibly may cause buckling, are known to exist.[19]

In order to demonstrate the *elastic* stress pattern that might constitute the prebuckled stress state, the shell was analyzed for an internal pressure of 150 psi. The use of the prebuckled stress state in a linear buckling analysis is described in Section 5.6.1. However, for torospherical shells, a

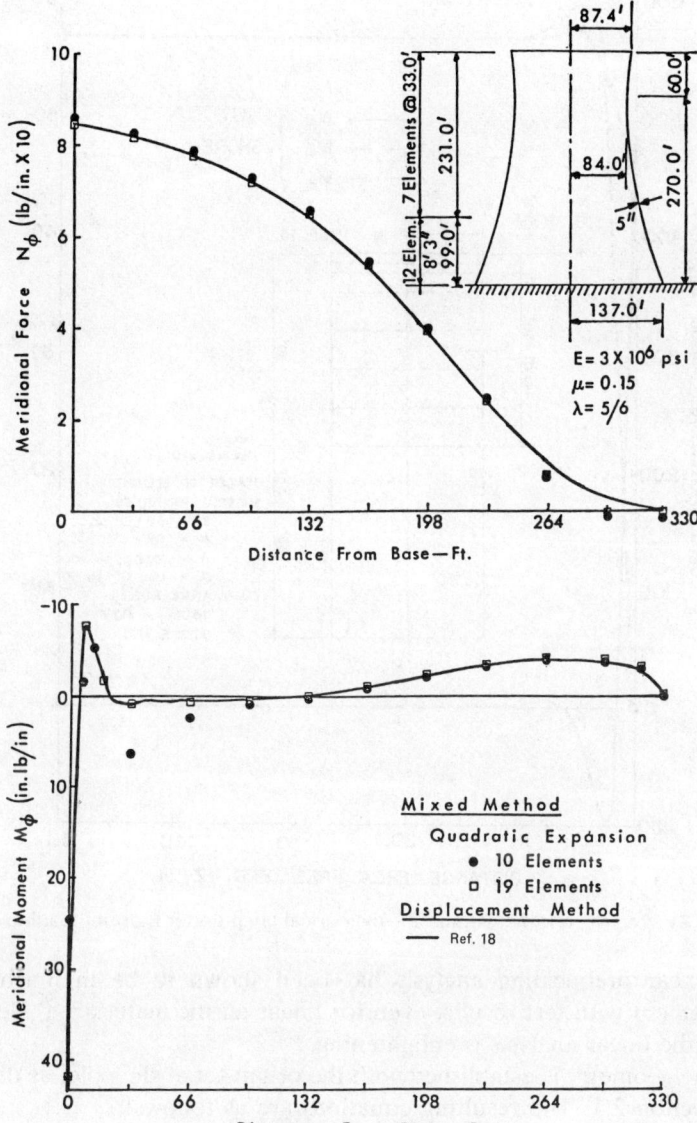

Fig. 3.20 Hyperboloidal shell under static wind loading

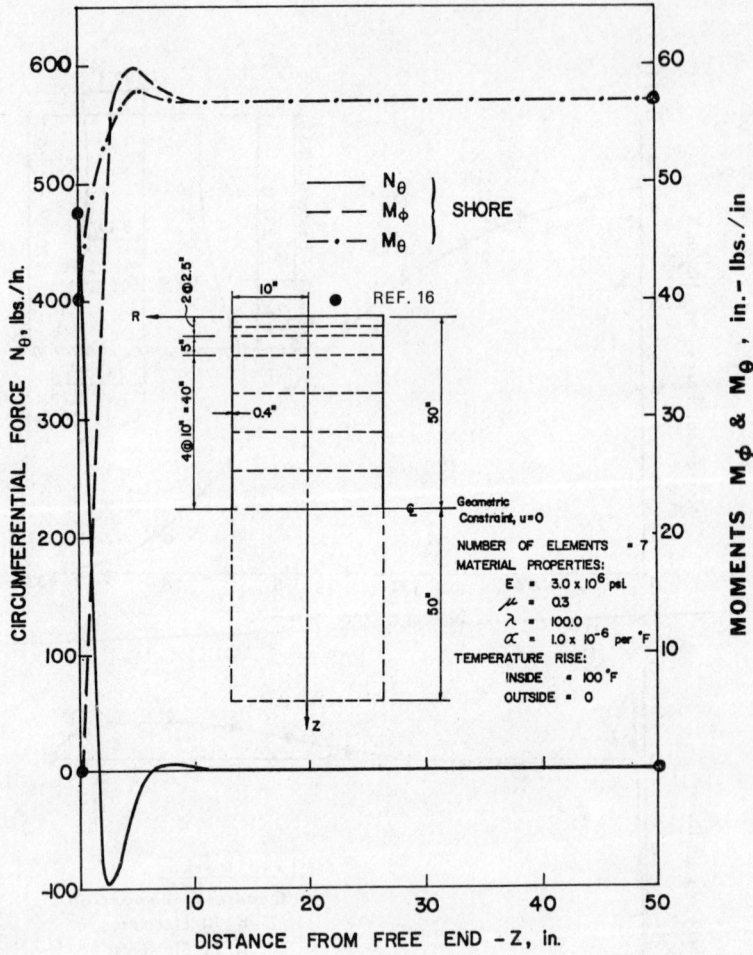

Fig. 3.21 N_θ, M_ϕ and M_θ versus Z—cylindrical shell under thermal loading

nonlinear prebuckling analysis has been shown to be in much better agreement with test results, even for linear elastic materials.[20] Nevertheless, the linear analysis is enlightening.

The geometry is established with the origin set at the pole, as discussed in Section 2.1. The resulting equations are as follows:

Spherical cap: $\quad Z^2 + R^2 - 345.8Z = 0$

Toroidal knuckle: $\quad Z^2 + R^2 - 95.7574Z - 126.72R + 5234.9519 = 0$

Cylindrical segment: $\quad R - 96.1 = 0$

The pertinent material properties are $E = 30\,000$ psi and $\nu = 0.3$.

Static analysis

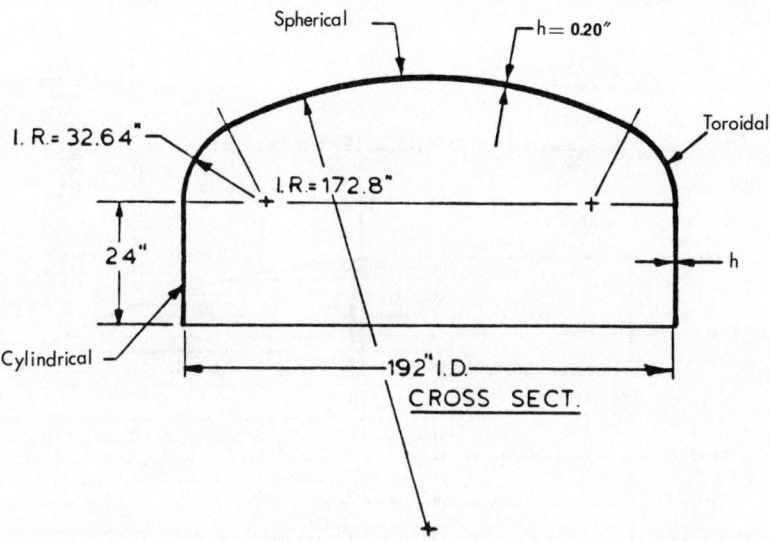

Fig. 3.22(a) Geometry of cylindrical shell with torospherical head

Fig. 3.22(b) Test specimen

Finite element analysis of shells of revolution

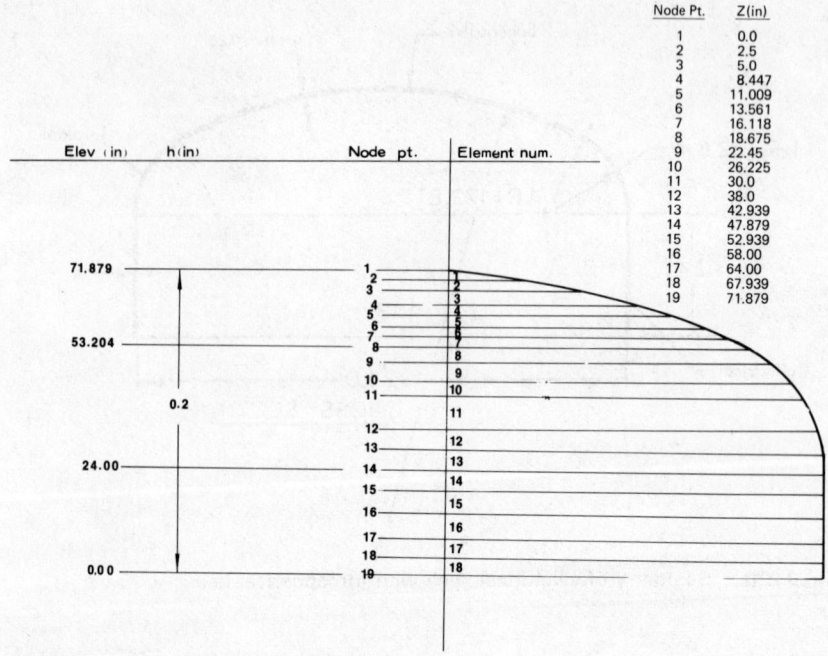

Fig. 3.23 Finite element discretization of shell

The finite element discretization is shown on Fig. 3.23 and the results of the analysis are given on Fig. 3.24 for the meridional stresses σ_ϕ, and on Fig. 3.25 for the circumferential stresses σ_θ. Total stresses (extensional plus bending components) at the inner and outer surfaces are shown, instead of stress resultants and couples, to demonstrate the local influence of bending in the junction regions. The plots represent the SHORE-3 solution and the triangles are results obtained using the BOSOR program,[21] which is discussed in detail in Section 7.2.1. The discrepancies in the cylindrical segment are due to slightly different boundary conditions in the BOSOR model. A further confirmation was obtained using the program ROT B,[22] which produced practically identical results to SHORE-3. The high circumferential compression that is characteristic of torospherical heads is clearly demonstrated on Fig. 3.25.

To illustrate the influence of bending on the extensional stresses, the circumferential stress σ_θ at the middle surface is shown on Fig. 3.26. A (hypothetical) second case, where both the thickness and the pressure are reduced to 10% of the original values, was analyzed, and the results are given along with the comparable stresses from a membrane theory

Static analysis

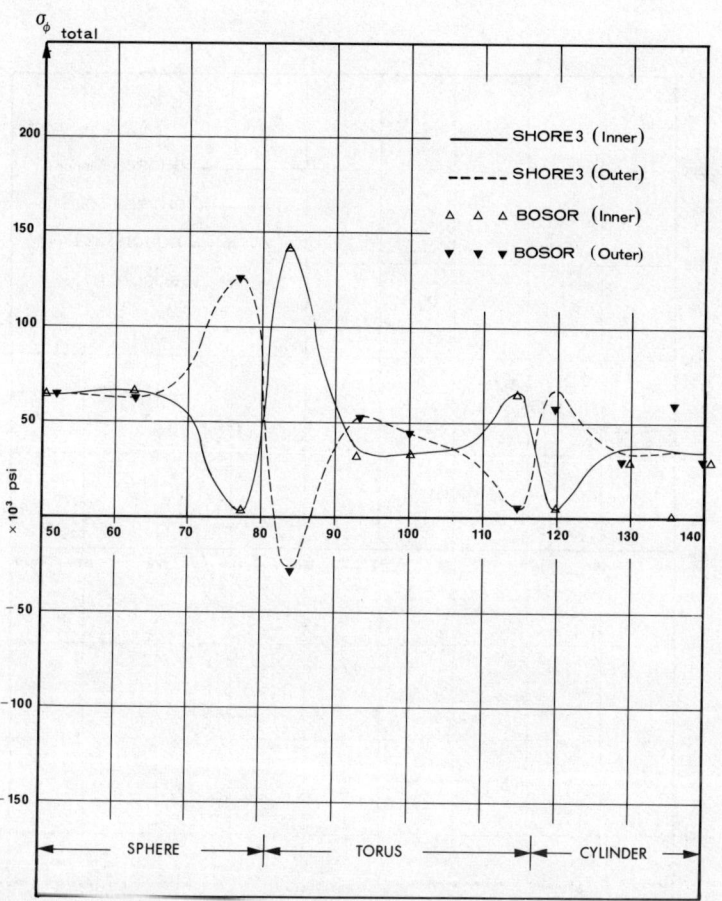

Fig. 3.24 Total meridional stresses on inner and outer surfaces

Finite element analysis of shells of revolution

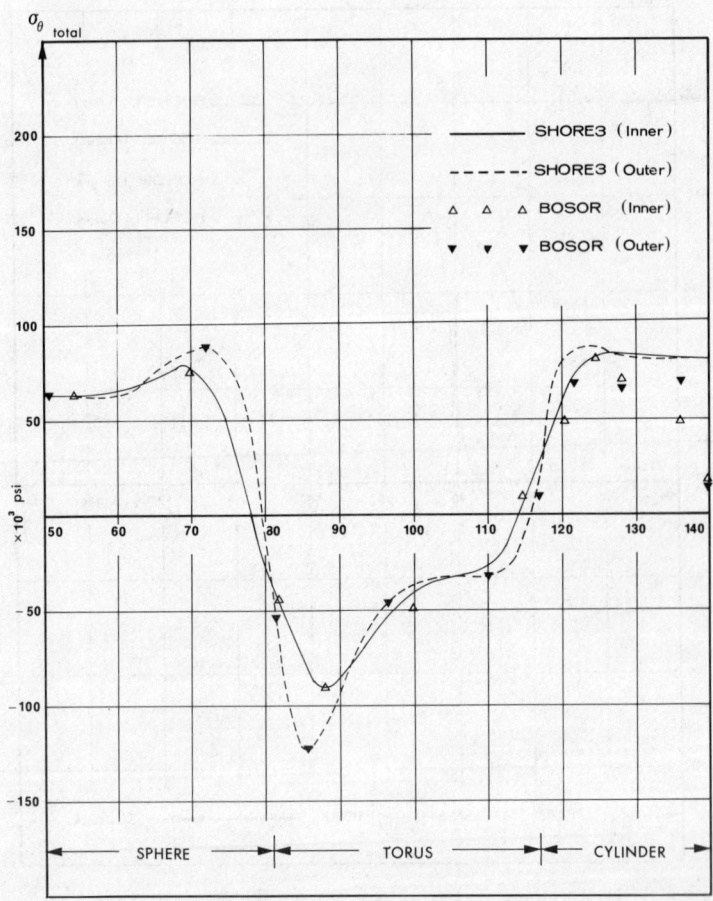

Fig. 3.25 Total circumferential stresses on inner and outer surfaces

Static analysis

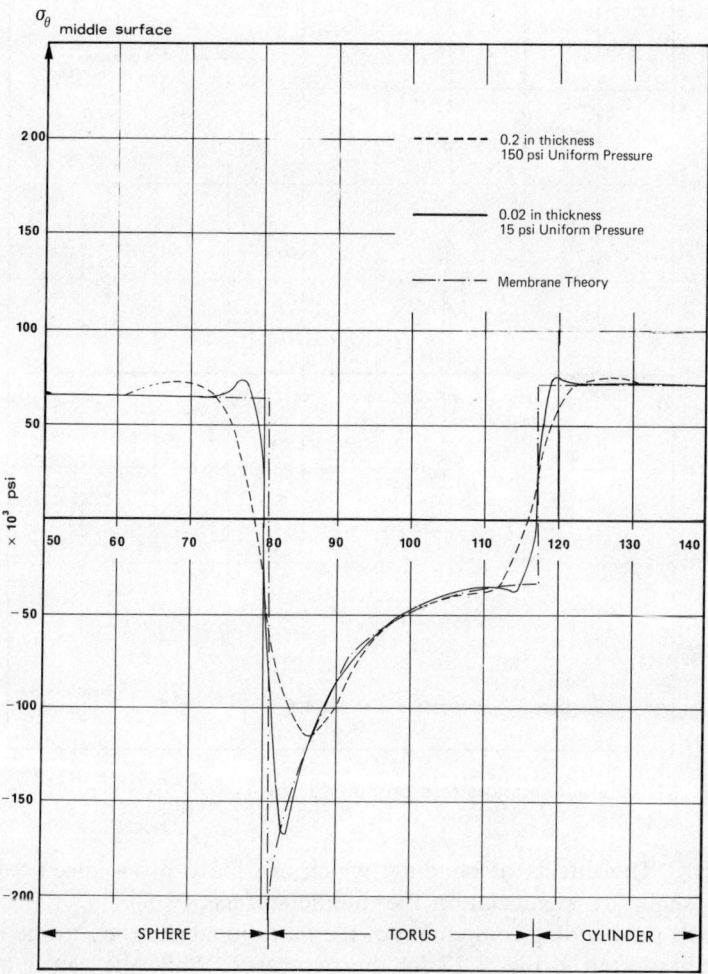

Fig. 3.26 Circumferential stress on middle surface

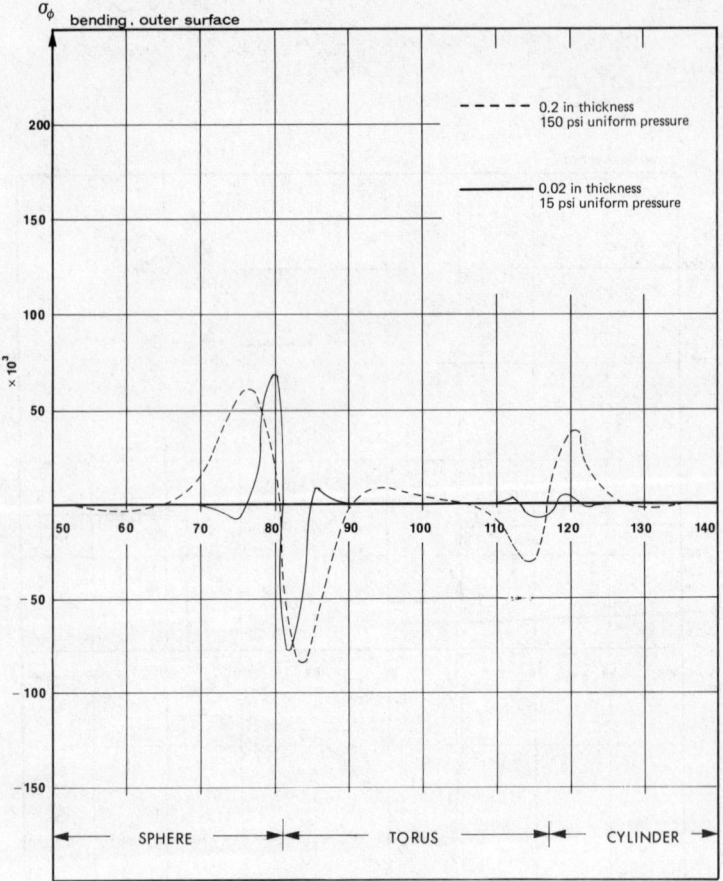

Fig. 3.27 Meridional bending stress on outer surface

analysis.[23] The effects of bending, which are more pronounced for the thicker shell, are beneficial on the middle surface.

Finally, the bending component of the meridional stress σ_ϕ at the outer surface is plotted in Fig. 3.27 for the two cases. While the values at the spherical–toroidal junction are not markedly different, the bending stresses within the toroidal region almost vanish for the thinner shell. At the toroidal–cylindrical junction, the thicker shell produces slightly larger bending stresses.

This analysis demonstrates several interesting aspects of the significance of bending stresses for a situation where there are neither abrupt geometrical discontinuities nor locally concentrated loads, two classical sources of bending in thin shells.[19]

References

1. Stricklin, J. A., 'Geometrically Nonlinear Static and Dynamic Analysis of Shells of Revolution', IUTAM Symposium on High Speed Computing of Elastic Structures, Liege, Belgium, Aug. 1970.
2. Basu, P. K. and Gould, P. L., 'Dynamic and Nonlinear Analysis of Shells of Revolution', Research Report No. 46, Structural Div., Dept. of Civil Engineering, Washington University, St. Louis, Mo., April 1977.
3. Basu, P. K. and Gould, P. L., 'Finite Element Discretization of Open-Type Axisymmetric Elements', *Int. J. Num. Methods in Engrg*, Vol. 14, 1979, pp. 159–178.
4. Fonder, G. A. and Clough, R. W., 'Explicit Addition of Rigid-Body Motions in Curved Finite Elements', *AIAA J.*, Vol. 11, No. 3, March 1973, pp. 305–312.
5. Mebane, P. M. and Stricklin, J. A., 'Implicit Rigid Body Motion in Curved Finite Elements', *AIAA J.*, Vol. 9, No. 2, Feb. 1971, pp. 344–345.
6. Bushnell, D., 'Computerized Analysis of Shells-Governing Equations', *J. Computers and Structures*, Vol. 18, No. 3, pp. 471–536.
7. Buchert, K. P., 'Buckling of Shell and Shell-Like Structures', K. P. Buchert and Assoc., Columbia, Mo., 1973, pp. 5–10, 31–34.
8. Yu, W. W., *Cold-Formed Steel Structures*, McGraw-Hill Book Co., New York, 1973, Ch. 3.
9. Gere, J. M. and Weaver, W. Jr., *Analysis of Framed Structures*, Van Nostrand and Reinhold Inc., New York, 1965, p. 198.
10. Aas-Jakobsen, A., 'Beitrag zur Theorie der Kugelschale auf Einzelstützen', *Ingr. Arch*, 1937, Vol. 7, pp. 275–294.
11. Gould, P. L. and Lee, S. L., 'Column-Supported Hyperboloids under Wind Load', *Publ. IABSE*, Zurich, 1971, pp. 47–64.
12. El-Shafee, O. M. and Gould, P. L., 'Local Stress Analysis of Discretely Supported Rotational Shells: Method and Computer Program', *Engineering Structures*, Vol. 1, 1979, pp. 153–161.
13. Gould, P. L. and Sen, S. K., 'Refined Mixed Method Finite Elements for Shells of Revolution', *Proc. Air Force Third Conf. on Matrix Methods in Struct. Mechanics*, Wright-Patterson AFB, Ohio, Oct. 1971, AFFOL-TR-77-160, Dec. 1973, pp. 397–421.
14. Szabo, B. A., 'Some Recent Developments in Finite Element Analysis', *Computers and Mathematics with Applications*, Vol. 5, 1979, pp. 99–115.
15. Sen, S. K. and Gould, P. L., 'Criteria for Finite Element Discretization of Shells of Revolution', *Int. J. Num. Methods in Engrg.*, Vol. 6, 1973. pp. 265–274.
16. Timoshenko, S. and Woinowski-Kreiger, S., *Theory of Plates and Shells*, 2nd ed., McGraw-Hill, New York, 1959, pp. 466–481.
17. Witmer, E. A., 'Discrete-Element Analysis of Shells of Revolution', Lecture Notes for Summer Program 1.595 on 'Finite Element Methods in Solid Mechanics', Massachusetts Institute of Technology, Cambridge, Mass., June 1968.

18. Hill, D. W. and Coffin, G. K., 'Stresses and Deflections in Cooling Tower Shells Due to Wind Loadings', *Bulletin of the IASS*, No. 35, Sept. 1968, pp. 45–51.
19. Gould, P. L., *'Static Analysis of Shells,'* Lexington Press, 1977, pp. 70–71, 102–106, 348–349.
20. Bushnell, D., 'Nonsymmetric Buckling of Internally Pressurized Ellipsoidal and Torospherical Elastic-Plastic Pressure Vessel Heads, *J. Pressure Vessel Technology, ASME,* Vol. 99, Feb. 1977, pp. 54–63.
21. Mokhtarian, K., 'Private Communication'.

    For information on BOSOR, see Bushnell, D., 'Stresses, Stability and Vibration of Complex Branched Shells of Revolution', *J. Computers and Structures,* Vol. 4, 1974, pp. 399–435.
22. Wunderlich, W., 'Private Communication'.

    For information on ROT B, see Wunderlich, W., Rensch, H. J. and Obrecht, H., 'Analysis of Elastic-Plastic Buckling and Imperfection Sensitivity of Shells of Revolution in Buckling of Shells' (E. Ramm, Editor), *Proc. of a State-of-the-Art Colloquium,* Springer-Verlag, Berlin, 1982, pp. 137–174.
23. Rotter, M., 'Private Communication'.

4 Dynamic analysis

4.1 Element equations of motion

4.1.1 Uncondensed equations

The dynamic analysis of rotational shells is based on Eq. 2.59, Hamilton's Variational Principle.

The variation of the functional $\delta \bar{L}$ is evaluated for element i and harmonic j in a similar manner as described for the static case in Section 3.2.1. One additional step is to relate the displacement vector $\{q\}$ to the reference surface displacement vector $\{D\}$, defined in Eq. 2.12, by

$$\{q_i(\zeta)\} = \{q_\phi q_\theta q_\zeta\} = [G(\zeta)]\{D_i\} \tag{4.1}$$

in which

$$[G(\zeta)] = \begin{bmatrix} 1 & 0 & 0 & \zeta & 0 \\ 0 & 1 & 0 & 0 & \zeta \\ 0 & 0 & 1 & 0 & 0 \end{bmatrix} \quad -\frac{h}{2} \le \zeta \le \frac{h}{2} \tag{4.2}$$

which reflects the basic assumption of linear shell theory regarding the normal to the middle surface remaining straight and normal during deformation. Also, for completeness, viscous damping proportional to the instantaneous velocity is included and, for simplicity, the thermal load is omitted.

With the preceding generalizations, the element equation analogous to Eq. 3.29 is

$$[k_i^j]\{\bar{Y}_i^j\} + [c_i^j]\{\dot{\bar{Y}}_i^j\} + [m_i^j]\{\ddot{\bar{Y}}_i^j\} = \{\bar{f}_i^j(t)\} + \{\bar{F}_i^j(t)\} \tag{4.3}$$

In Eq. 4.3, $[k_i^j]$ = the element stiffness matrix previously defined in Eq. 3.25. Additionally,

$$[m_i^j] = L_i \lambda \pi \int_0^1 \int_{-h/2}^{h/2} \rho [\bar{B}]^T [G]^T [G] [\bar{B}] R \, d\zeta \, ds \tag{4.4}$$

$\quad$ = element (consistent) mass matrix

and

$$[c_i^j] = \text{Element damping matrix} \quad (4.5)$$

With respect to Eqs 4.4 and 4.5, $[\bar{B}]$ is given by Eq. 3.23(b) and ρ was defined in Section 2.7.1; the form of $[c_i^j]$ will be specified at the global level.

4.1.2 Condensed equations

Analogous to Eq. 3.30, Eq. 4.3 is written in partitioned form as

$$\begin{bmatrix} \mathbf{k}_{iaa}^j & \mathbf{k}_{iab}^j \\ \mathbf{k}_{iba}^j & \mathbf{k}_{ibb}^j \end{bmatrix} \begin{Bmatrix} \Delta_i^j \\ \bar{\mathbf{Y}}_{ib}^j \end{Bmatrix} + \begin{bmatrix} \mathbf{m}_{iaa}^j & \mathbf{m}_{iab}^j \\ \mathbf{m}_{iba}^j & \mathbf{m}_{ibb}^j \end{bmatrix} \begin{Bmatrix} \ddot{\Delta}_i^j \\ \ddot{\bar{\mathbf{Y}}}_{ib}^j \end{Bmatrix} = \begin{Bmatrix} \bar{\mathbf{f}}_{ia}^j \\ \bar{\mathbf{f}}_{ib}^j \end{Bmatrix} + \begin{Bmatrix} \bar{\mathbf{F}}_i^j \\ \mathbf{0} \end{Bmatrix} \quad (4.6)$$

where, for the moment, damping is neglected. Then, Eq. 4.6 is expanded into

$$\mathbf{k}_{iaa}^j \Delta_i^j + \mathbf{k}_{iab}^j \bar{\mathbf{Y}}_{ib}^j + \mathbf{m}_{iaa}^j \ddot{\Delta}_i^j + \mathbf{m}_{iab}^j \ddot{\bar{\mathbf{Y}}}_{ib}^j = \bar{\mathbf{f}}_{ia}^j + \bar{\mathbf{F}}_i^j$$

$$\mathbf{k}_{iba}^j \Delta_i^j + \mathbf{k}_{ibb}^j \bar{\mathbf{Y}}_{ib}^j + \mathbf{m}_{iba}^j \ddot{\Delta}_i^j + \mathbf{m}_{ibb}^j \ddot{\bar{\mathbf{Y}}}_{ib}^j = \bar{\mathbf{f}}_{ib}^j + \mathbf{0} \quad (4.7)$$

which is similar to Eq. 3.31, with the addition of the time-dependent terms.

A further reduction of the equations is accomplished by using the *kinetic condensation* procedure as derived by Geradin[1] and applied to the case of a shell of revolution by Sen and Gould.[2] Essentially, assuming simple harmonic motion and suppressing frequency dependent higher order terms, the condensed undamped equations of motion are approximated as

$$[\bar{k}_i^j]\{\Delta_i^j\} + [\bar{m}_i^j]\{\ddot{\Delta}_i^j\} = \{\bar{\mathcal{F}}_i^j\} \quad (4.8)$$

in which $[\bar{k}_i^j]$ has been previously defined in Eq. 3.34 and

$$[\bar{m}_i^j] = [\mathbf{m}_{iaa}^j + \mathbf{k}_{iab}^j \mathbf{k}_{ibb}^{j-1} \mathbf{m}_{ibb}^j \mathbf{k}_{ibb}^{j-1} \mathbf{k}_{iba}^j - \mathbf{k}_{iab}^j \mathbf{k}_{ibb}^{j-1} \mathbf{m}_{iba}^j - \mathbf{m}_{iab}^j \mathbf{k}_{ibb}^{j-1} \mathbf{k}_{iba}^j] \quad (4.9)$$

Also, omitting the thermal terms,

$$\{\bar{\mathcal{F}}_i^j\} = \{\bar{\mathbf{f}}_{ia}^j + \bar{\mathbf{F}}_i^j - \mathbf{k}_{iab}^j \mathbf{k}_{ibb}^{j-1} \bar{\mathbf{f}}_{ib}^j\} \quad (4.10)$$

In contrast to static condensation, the kinematic counterpart is not exact since frequency dependent terms are omitted. This is discussed further in Section 4.4.1.

4.2 Global equations of motion

Following the assembly procedure introduced in Section 3.2.3.1; using the uncondensed element equations Eq. 4.8; applying the boundary conditions; and reintroducing viscous damping, the global equations of motion are

$$[\bar{K}^j]\{\Delta^j\}+[\bar{C}^j]\{\dot{\Delta}^j\}+[\bar{M}^j]\{\ddot{\Delta}^j\}=\{\bar{\mathscr{F}}^j(t)\} \quad (4.11)$$

in which $[\bar{K}^j]$, $[\bar{C}^j]$ and $[\bar{M}^j]$ are the global stiffness, damping and consistent mass matrices and $\{\bar{\mathscr{F}}^j\}$ is the global consistent load vector. There are a total of $5(N+1)$ equations where $N=$ the number of finite elements in the model.

$[\bar{M}^j]$ is a variationally consistent mass matrix derived using the same interpolation functions as employed for the stiffness matrix and load vector. This form of the mass matrix is relatively tightly banded, but non-diagonal. While this characteristic can be somewhat restrictive, particularly in direct integration solutions, it is felt that advantages attained in accuracy and uniformity dictate the exclusive use of consistent mass matrices for this class of problems.

In order to handle the somewhat elusive damping contribution conveniently, it is common to assume *proportional damping*, whereby the damping matrix is taken as a linear combination of the stiffness and mass matrices, i.e.,

$$[\bar{C}^j]=\alpha_1[\bar{M}^j]+\alpha_2[\bar{K}^j] \quad (4.12)$$

in which the constant multipliers α_1 and α_2 are related to the damping ratio (percent of critical damping) β_r^j for any meridional mode r in harmonic j by

$$\beta_r^j=\frac{\alpha_1}{2\omega_r^j}+\frac{\alpha_2}{2}\omega_r^j \quad (4.13)$$

where ω_r^j is the corresponding frequency (see Section 4.4.1). This is sufficiently general to provide a given damping ratio at two selected frequencies. Alternately, it may be more convenient to specify $\beta_r^j=\bar{\beta}$ for a given frequency $\omega_r^j=\bar{\omega}$ on the basis of test data. Then, with

$$\alpha_1=\bar{\beta}\bar{\omega} \quad (4.14)$$

and

$$\alpha_2=\bar{\beta}/\bar{\omega} \quad (4.15)$$

$[\bar{C}^j]$, as given by Eq. 4.12, may be evaluated.

4.3 Extensions to basic element

4.3.1 General

The preceding rotational shell formulation including dynamic terms indicates that any elaborations on the basic element, e.g. stiffened shells, branched shells, open-type elements will require only the addition of an appropriate mass matrix.

For the case of stiffened shells it is probably sufficient to spread the mass of the stiffeners uniformly over the tributary shell region, while the consistent mass for the open-type element can be developed by using the Hermitian interpolation functions introduced in Section 3.3.5.2 to evaluate the stiffness matrices.[3]

4.3.2 Consistent mass matrix for open-type element

4.3.2.1 Member mass matrix

Using the displacement functions Eqs 3.62 and 3.63, the consistent mass matrix for a prismatic member is given by

$$[\hat{m}] = \int_V [Z]^T [\hat{\rho}][Z] \, dV \qquad (4.16)$$

where

$$[\hat{\rho}] = \begin{bmatrix} \rho & & & 0 \\ & \rho & & \\ & & \rho & \\ 0 & & & \rho r^2 \end{bmatrix} \qquad (4.17)$$

and

$r^2 = y^2 + z^2$, the radial distance of a point from the centroidal axis.

After carrying out the integrations with respect to y and z, the mass matrix becomes

$$[\hat{m}] = \int_0^{L_c} [Z]^T [\hat{D}][Z] \, dx \qquad (4.18)$$

in which

$$[\hat{D}] = \begin{bmatrix} \rho A & & & 0 \\ & \rho A & & \\ & & \rho A & \\ 0 & & & \rho I_p \end{bmatrix} \qquad (4.19)$$

A = cross-sectional area of the section

and

I_p = polar moment of inertia of the section = $I_y + I_z$.

Dynamic analysis

Upon substituting $[\hat{D}]$ and $[Z]$ from Eqs 4.19 and 3.63(d), respectively, into Eq. 4.18 and carrying out the necessary integrations, the resulting form of the mass matrix for a member, with respect to the local coordinate system, will be:

$[\hat{m}] =$

$$\begin{bmatrix}
\frac{A\rho L_c}{3} & 0 & 0 & 0 & 0 & 0 & \frac{A\rho L_c}{6} & 0 & 0 & 0 & 0 & 0 \\
 & \frac{13A\rho L_c}{35} & 0 & 0 & 0 & -\frac{11A\rho L_c^2}{210} & 0 & \frac{9A\rho L_c}{70} & 0 & 0 & 0 & \frac{13A\rho L_c^2}{420} \\
 & & \frac{13A\rho L_c}{35} & 0 & \frac{11A\rho L_c^2}{210} & 0 & 0 & 0 & \frac{9A\rho L_c}{70} & 0 & -\frac{13A\rho L_c^2}{420} & 0 \\
 & & & \frac{I_p\rho L_c}{3} & 0 & 0 & 0 & 0 & 0 & \frac{I_p\rho L_c}{6} & 0 & 0 \\
 & & & & \frac{A\rho L_c^3}{105} & 0 & 0 & 0 & \frac{13A\rho L_c^2}{420} & 0 & -\frac{A\rho L_c^3}{140} & 0 \\
 & & & & & \frac{A\rho L_c^3}{105} & 0 & -\frac{13A\rho L_c^2}{420} & 0 & 0 & 0 & -\frac{A\rho L_c^3}{140} \\
 & & & & & & \frac{A\rho L_c}{3} & 0 & 0 & 0 & 0 & 0 \\
 & \text{(Symmetric)} & & & & & & \frac{13A\rho L_c}{35} & 0 & 0 & 0 & -\frac{11A\rho L_c^2}{210} \\
 & & & & & & & & \frac{13A\rho L_c}{35} & 0 & \frac{11A\rho L_c^2}{210} & 0 \\
 & & & & & & & & & \frac{I_p\rho L_c}{3} & 0 & 0 \\
 & & & & & & & & & & \frac{A\rho L_c^3}{105} & 0 \\
 & & & & & & & & & & & \frac{A\rho L_c^3}{105}
\end{bmatrix}$$

(4.20)

4.3.2.2 Coordinate transformations

Pursuant to Section 3.3.5.4, the mass matrix transformation from local to curvilinear coordinates is given by

$$[\hat{m}_1] = [T_1]^T [\hat{m}][T_1] \tag{4.21}$$

and the further meridional slope transformation is expressed by

$$[\hat{m}_2] = [T_2]^T [\hat{m}_1][T_2] \tag{4.22}$$

If no slope discontinuity is present, $[\hat{m}_2] = [\hat{m}_1]$.

The condensation to remove the degrees of freedom corresponding to the rotation about the normal is

$$[\hat{m}'] = [\hat{m}_{rr}] + [\hat{k}_{rc}][\hat{k}_{cc}]^{-1}([\hat{m}_{cc}][\hat{k}_{cc}]^{-1}[\hat{k}_{cr}] - [\hat{m}_{cr}]) - [\hat{m}_{rc}][\hat{k}_{cc}]^{-1}[\hat{k}_{cr}] \tag{4.23}$$

where the elements of $[\hat{k}_2]$ have been defined in Eq. 3.70 and where $[\hat{m}_2]$ is partitioned identically.

As noted earlier, no condensation is necessary if the open-type members have full continuity with the adjacent shell elements so that

$$[\hat{m}'] = [\hat{m}_{rr}] \tag{4.24}$$

4.3.2.3 Effective mass matrix

As elaborated in Section 3.3.5.6 the effective mass matrix is derived by smearing the member properties over their tributary arc and summing the effect of all members around the circumference.[3] The elements of the effective mass matrix $[m]$ are found from the elements of $[\hat{m}']$, just as in Eq. 3.74.

For harmonic j, the final form of the mass matrix for an open-type element is

$$[\bar{m}^j] = \frac{1}{f} \int_{-\pi}^{\pi} \lceil \tilde{\Theta} \rfloor [m] \lceil \tilde{\Theta} \rfloor \, d\theta \tag{4.25}$$

where $\lceil \tilde{\Theta} \rfloor$ and f were defined in the earlier section. The element mass matrices are then assembled into the global mass matrix $[\bar{M}^j]$, in the same manner as $[\bar{K}^j]$, as described in Section 3.2.3.1.

4.4 Modal superposition solutions

4.4.1 Free vibration

Perhaps the most incisive approach to the solution of problems in structural dynamics is to assume the motion to be harmonic with frequencies ω. This is expressed by

$$\{\Delta^j(t)\} = \{\bar{\Delta}^j\} e^{i\omega t} \tag{4.26}$$

Introducing Eq. 4.26 into the homogeneous part of Eq. 4.11 and neglecting damping yields

$$([\bar{K}^j] - \omega^2[\bar{M}^j])\{\bar{\Delta}^j\} = \{0\} \tag{4.27}$$

Equation 4.27 is called the free vibration equation and is a linear eigenvalue problem. The eigenvalues ω_r^j are the *circular natural frequencies*, and the eigenvectors $\{\bar{\Delta}_r^j\}$ are the corresponding meridional mode shapes in circumferential harmonic j, up to a maximum of $\bar{r}$ per harmonic. $\bar{r}$ is generally set based on the frequency content of the anticipated forcing function. For systems with low to moderate damping, which includes most thin shells, the undamped solution is sufficiently accurate for free vibration analysis.

The solution of the linear eigenvalue problem, may become somewhat involved, especially for systems with a large number of coordinates, but has been well documented. For the class of problems under consideration here, the EASI modification of the Sturm sequence method put forth by Gupta[4] has proved to be effective.

In assessing the quality of any solution to Eq. 4.27, it should be recalled that the stiffness and mass matrices will likely have been reduced to manageable size by kinematic condensation. As briefly noted in Section 4.1.2, this procedure is only approximate; thus some examination of the magnitude of the error introduced is in order.

In Ref. 2, it was shown that the accuracy of the eigenvalues ω^2 calculated using condensed equations of motion *decreases* as the order of approximation function selected for the displacements *increases*. This result, which may seem surprising initially, seems to conflict with the notion of convergence with increasing order of polynomial approximation. It should be emphasized however, that this applies only to condensed equations; uncondensed systems will yield improved results with increased order of approximation. However, it is often impractical to use uncondensed matrices because of the tremendous increase in degrees of freedom (DOF).

To illustrate this point, the errors in natural frequencies for three representative shells, as compared to results obtained from analytical solutions, are given in Table 4.1. It is apparent that the calculated frequencies using condensed equations are always higher than the reference values and that the error increases with the order of approximation

Table 4.1

Shell type	j	r mode	No. of elements	ω_r^i (rad/sec) Reference solutions	Percentage error in ω_r^i, $n =$ order of comparison function			
					3	4	6	Variable
Cylindrical	0	1	12	33 440	2.61	3.35	3.35	3.08
		2		49 640	1.95	7.76	7.84	1.97
		3		50 580	0.91	6.09	6.25	0.93
Hemispherical	1	1	11	0.5417	0.13	0.42	0.42	0.13
		2		0.8523	0.76	3.04	3.06	0.77
Hyperboloidal	1	1	11	20.660	0.01	0.63	0.63	0.01
		2		42.350	1.16	4.27	4.42	1.18
		3		66.100	1.39	10.24	14.17	2.45
	5	1	6	6.502	0.06	0.37	0.42	0.08
		2		8.981	0.36	0.86	0.94	0.23
		3		12.920	0.39	1.86	2.09	0.31

n. The reason for decreasing accuracy with higher order of approximation is that each condensed DOF corresponds to the imposition of a constraint which tends to artificially stiffen the system.[2]

An obvious remedy would be to use cubic or even lower order displacement functions. With low-order approximations, however, the predictions of stresses in the shell tends to be erratic, particularly in regions of steep gradients. The best compromise may be the use of variable order elements, quadratic or cubic, except in expected regions of stress fluctuation where fourth to sixth order elements should be employed. Results obtained by the use of such a scheme are given in the last column of Table 4.1 and seem to be acceptable.

Also, with regard to the eventual evaluation of stresses in the system, it should be noted that seemingly small differences in natural frequency can be magnified in the calculation of the mode shapes and, ultimately, the stresses. Therefore, every effort should be made to compute the frequencies as precisely as practical.

4.4.2 Generalized coordinates

Since the free vibration solution provides the homogeneous part of Eq. 4.11, the complete dynamic solution requires only a particular solution. This is best accomplished by a coordinate transformation

$$\{\Delta^j(t)\} = [\bar{\Delta}^j]\{\eta^j(t)\} \tag{4.28}$$

in which

$$[\bar{\Delta}^j] = [\{\bar{\Delta}_1^j\}\{\bar{\Delta}_2^j\} \ldots \{\bar{\Delta}_r^j\} \ldots \{\bar{\Delta}_{\bar{r}}^j\}] \tag{4.29}$$

and = matrix of mode shape vectors found from the solution of Eq. 4.27, with $\bar{\Delta}_r^j$ corresponding to ω_r^j, etc.

$$\{\eta^j(t)\} = \{\eta_1^j(t)\ \eta_2^j(t) \ldots \eta_r^j(t) \ldots \eta_{\bar{r}}^j(t)\} \tag{4.30}$$

= vector of generalized coordinates

The introduction of Eq. 4.28 into Eq. 4.11 and the use of the orthogonality properties of the mass and stiffness matrices[5] yields an *uncoupled* set of equations in the generalized coordinates

$$\sum_{r=1}^{\bar{r}} \ddot{\eta}_r^j + 2\beta_r^j \omega_r^j \dot{\eta}_r^j + (\omega_r^j)^2 \eta_r^j = \bar{p}_r^j/\bar{m}_r^j \tag{4.31}$$

in which

$$\omega_r^j = (\bar{k}_r^j/\bar{m}_r^j)^{1/2} \tag{4.32a}$$

and = circular natural frequency for mode r of harmonic j

$$\beta_r^j = \bar{c}_r^j/(2\omega_r^j \bar{m}_r^j) \tag{4.32b}$$

= damping ratio for mode r of harmonic j.

Also,

$$\bar{k}_r^j = \lfloor\bar{\Delta}_r^j\rfloor[\bar{K}^j]\{\bar{\Delta}_r^j\} \tag{4.33a}$$

= term r of generalized stiffness $[\bar{\Delta}^j]^T[\bar{K}^j][\bar{\Delta}^j]$

$$\bar{c}_r^j = \lfloor\bar{\Delta}_r^j\rfloor[\bar{C}^j]\{\bar{\Delta}_r^j\} \tag{4.33b}$$

= term r of generalized damping $[\bar{\Delta}^j]^T[\bar{C}^j][\bar{\Delta}^j]$

$$\bar{m}_r^j = \lfloor\bar{\Delta}_r^j\rfloor[\bar{M}^j]\{\bar{\Delta}_r^j\} \tag{4.33c}$$

= term r of generalized mass $[\bar{\Delta}^j]^T[\bar{M}^j][\bar{\Delta}^j]$

$$\bar{p}_r^j = \lfloor\bar{\Delta}_r^j\rfloor\{\bar{\mathscr{F}}^j\} \tag{4.33d}$$

= term r of generalized load $[\bar{\Delta}^j]^T\{\bar{\mathscr{F}}^j\}$

Equations 4.31 may be solved individually by the Duhamel Integral in the form[6]

$$\eta_r^j = \frac{1}{\bar{m}_r^j \omega_r^j} \int_0^t \bar{p}_r^j(\tau) e^{-\beta_r^j \omega_r^j (t-\tau)} \sin \omega_r^j(t-\tau)\, d\tau \quad (r = 1, \ldots, \bar{r}) \tag{4.34}$$

In the event that the initial displacement and/or velocity are not zero, a free vibration solution must be included. Since this modification is somewhat specialized, and not particularly germane to this class of problems, it is not developed here. The subject is discussed in Ref. 6, and in other books on structural dynamics.

Once the solution for the generalized coordinates is obtained, the time dependent displacement vector for each harmonic in curvilinear coordinates is found from Eq. 4.28.

4.4.3 Uniform base motion

4.4.3.1 Modification of equations of motion

It is common to represent an earthquake induced excitation of a shell as a uniform ground motion with displacement, velocity and acceleration $z_g(t)$, $\dot{z}_g(t)$ and $\ddot{z}_g(t)$, respectively. This problem may also be treated in generalized coordinates if coordinates relative to the base are used.[7] The formulation is restricted to $j = 0$ (vertical or axial base motion) and $j = 1$ (horizontal or transverse base motion) and is accomplished by taking

$$\{\Delta^j(t)\} = \begin{Bmatrix} \Delta_A^j \\ \hdashline \Delta_B^j \end{Bmatrix} = \begin{Bmatrix} Y_A^j \\ \hdashline Y_B^j \end{Bmatrix} + \begin{Bmatrix} X_A^j \\ \hdashline X_B^j \end{Bmatrix} \tag{4.35}$$

in which

$\{Y_A^j\}$, $\{Y_B^j\}$ = shell displacements *due* to uniform ground motion
$\{X_A^j\}$, $\{X_B^j\}$ = shell displacements *relative* to the ground motion

and A and B refer to coordinates above and at the base, respectively.

The corresponding equations of motion, partitioned in accordance with Eq. 4.35 and reduced by kinematic condensation, are

$$[\bar{K}^j_{AA}]\{X^j_A\}+[\bar{C}^j_{AA}]\{\dot{X}^j_A\}+[\bar{M}_{AA}]\{\ddot{X}^j_A\}=\{\bar{\bar{\mathscr{F}}}^j_A\} \quad (4.36)$$

in which

$$\{\bar{\bar{\mathscr{F}}}_A\}=-[\bar{K}^j_{AA}]\{Y^j_A\}-[\bar{K}^j_{AB}]\{Y^j_B\}-[\bar{C}^j_{AA}]\{\dot{Y}^j_A\}$$
$$-[\bar{C}^j_{AB}]\{\dot{Y}^j_B\}-[\bar{M}^j_{AA}]\{\ddot{Y}^j_A\}-[\bar{M}^j_{AB}]\{\ddot{Y}^j_B\} \quad (4.37)$$

and $[\bar{C}^j_{AA}]$ and $[\bar{C}^j_{AB}]$ are reduced viscous damping matrices.

In Eq. 4.37, the first two terms must cancel, since they correspond to rigid body motion. Further, the damping terms are generally small compared to the inertial terms and will be neglected. The remaining terms may be expressed as a function of the uniform ground acceleration $\ddot{z}_g$, through a transformation between the Cartesian and the curvilinear coordinates, as

$$\{\bar{\bar{\mathscr{F}}}_A\}=-([\bar{M}_{AA}]\{R^j_A\}-[\bar{M}^j_{AB}]\{R^j_B\})\ddot{z}^j_g$$
$$=\{\bar{M}_A\}\ddot{z}^j_g \quad (4.38)$$

For vertical ground motion, $j = 0$, only displacements u, w and β_ϕ need be retained. Referring to Fig. 2.1,

$$\{R^0_A\}=\{\sin\phi_1 \; -\cos\phi_1 \; 0 \ldots \sin\phi_i \; -\cos\phi_i \; 0 \ldots \sin\phi_N \; -\cos\phi_N \; 0\} \quad (4.39a)$$

$$\{R^0_B\}=\{\sin\phi_{N+1} \; -\cos\phi_{N+1} \; 0\} \quad (4.39b)$$

where N = total number of elements.

For horizontal ground motion, $j = 1$, all five displacement components are present so that

$$\{R^1_A\}=\{-\cos\theta\cos\phi_1\sin\theta \; -\cos\theta\sin\phi_1 \; 0 \; 0 \ldots$$
$$-\cos\theta\cos\phi_i\sin\theta \; -\cos\theta\sin\phi_i \; 0 \; 0 \ldots \quad (4.40a)$$
$$-\cos\theta\cos\phi_N\sin\theta \; -\cos\theta\sin\phi_N \; 0 \; 0\}$$

$$\{R^1_B\}=\{-\cos\theta\cos\phi_{N+1}\sin\theta \; -\cos\theta\sin\phi_{N+1} \; 0 \; 0\} \quad (4.40b)$$

The equations of motion transform to generalized coordinates, Eq. 4.31, by a procedure identical to that used in Section 4.4.2. Here,

$$\{X^j_A\}=[\bar{\Delta}^j_A]\{\eta^j(t)\} \quad (4.41)$$

where the mode shapes $[\bar{\Delta}^j_A]$ are contained in $[\bar{\Delta}^j]$, Eq. 4.29. For this case, the terms defined in Eqs 4.33 are replaced by

$$\bar{k}^j_r = \lfloor\bar{\Delta}^j_{Ar}\rfloor[\bar{K}^j_{AA}]\{\bar{\Delta}^j_{Ar}\} \quad (a) \qquad \bar{m}^j_r = \lfloor\bar{\Delta}^j_{Ar}\rfloor[\bar{M}^j_{AA}]\{\bar{\Delta}^j_{Ar}\} \quad (c)$$
$$\bar{c}^j_r = \lfloor\bar{\Delta}^j_{Ar}\rfloor[\bar{C}^j_{AA}]\{\bar{\Delta}^j_{Ar}\} \quad (b) \qquad \bar{p}^j_r = \lfloor\bar{\Delta}^j_{Ar}\rfloor\{\bar{\bar{\mathscr{F}}}^j_A\} \quad (d) \quad (4.42)$$

4.4.3.2 Response spectrum analysis

It is common to represent an earthquake excitation by a response spectrum. Such a spectrum is a graph of the *maximum* value of a particular function, e.g. displacement, velocity or acceleration, when a single degree of freedom (SDF) system is subjected to a real or simulated time history base acceleration, corresponding to the selected earthquake. The natural frequencies and damping ratios are individually incremented through the range of interest to generate points on the graph. In this way, *time* is removed as an explicit variable, since only the maximum ordinate is plotted for each value of frequency and damping. A representative acceleration spectrum is shown in Fig. 4.1, in terms of the period T and damping ratio β.

For more general use, a spectrum may be scaled based on the ground acceleration, S_A for $T = 0$, which is about 11 ft/sec² or 0.33g for this plot. For example, in a less severe seismic region, 0.20g ground acceleration may be appropriate, so that the ordinates of Fig. 4.1 would be reduced by 0.2/0.33.

For the shell of revolution, the transformation to generalized coordinates enables a solution to be constructed as the superposition of the solutions of SDF systems. To do this, we first write

$$\{\bar{\bar{\mathscr{F}}}_A\}_{\max} = \{\bar{M}_A\}S^i_{\mathscr{A}r} = \omega^i_r\{\bar{M}_A\}S^i_{Vr} \qquad (4.43)$$

where $S^i_{\mathscr{A}r}$ is the acceleration response spectrum ordinate corresponding

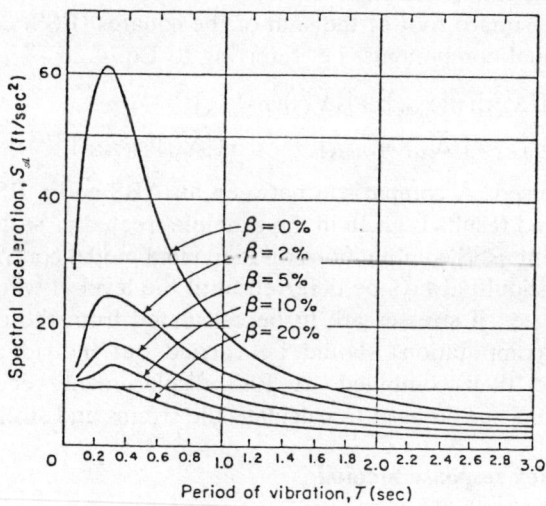

Fig. 4.1 Average acceleration response spectra, 1940 El Centro intensity. (From *U.S. Atomic Energy Commission Report TID-7024*, August 1963.)

to ω_r^j (or T_r^j) and β_r^j. Also,

$$S_{V_r}^i = \frac{1}{\omega_r^j} S_{\mathscr{A}r}^i$$

is the *pseudo-velocity spectrum* ordinate, which is commonly used in design.

Then, from Eqs 4.42(d) and 4.43,

$$(\bar{p}_r^j)_{\max} = \lfloor \bar{\Delta}_{Ar}^j \rfloor \{\bar{\bar{\mathscr{F}}}_A\}_{\max} = \omega_r^j \lfloor \bar{\Delta}_{Ar}^j \rfloor \{\bar{\bar{M}}_A\} S_{V_r}^i \qquad (4.44)$$

and, from Eq. 4.31,

$$(\eta_r^j)_{\max} = \frac{(\bar{p}_r^j)_{\max}}{\bar{m}_r^j(\omega_r^j)^2} = \frac{\lfloor \bar{\Delta}_{Ar}^j \rfloor \{\bar{\bar{M}}_A\}}{\bar{m}_r^j \omega_r^j} S_{V_r}^i \qquad (4.45)$$

The maximum relative displacement for mode r then follows from Eq. 4.41 as

$$\{X_{Ar}^j\}_{\max} = \{\bar{\Delta}_{Ar}^j\}(\eta_r^j)_{\max} \qquad (4.46)$$

Strains and stresses are introduced due to the relative displacements and may be computed as in Section 3.2.3.4.

To complete the solution, the preceding steps are repeated for all $\bar{r}$ participating meridional modes in circumferential harmonic j and substituted into Eq. 4.28. Before carrying this through, it is constructive to recall that the time dependence has been removed through the response spectrum approach. However, by the same token, the various values of $(\eta_r^j)_{\max}$ will likely occur at different times so that direct addition of absolute values (ABS) through Eq. 4.28 will probably overestimate the response. The square root of the sum of the squares (RSS) combinations of the individual components, i.e. referring to Eq. 4.28

$$\{X_A^j\}_{RSS} = \{[\{\bar{\Delta}_{A1}^j\}(\eta_1^j)_{\max}]^2 + [\{\bar{\Delta}_{A2}^j\}(\eta_2^j)_{\max}]^2$$
$$+ \ldots + [\{\bar{\Delta}_{Ar}^j\}(\eta_r^j)_{\max}]^2 + \ldots + [\{\bar{\Delta}_{A\bar{r}}^j\}(\eta_{\bar{r}}^j)_{\max}]^2\}^{1/2} \qquad (4.47)$$

is frequently used. A comparison between an ABS and a RSS combination for selected results is given in the example treated in Section 4.6.3.

In performing RSS combinations on individual modal contributions, the computation should always be deferred until the level at which the value will be used, i.e., if stresses are to be calculated from the displacement vectors, the computations should be carried out individually in each harmonic and then combined. A RSS displacement vector, such as $\{X_A^j\}_{RSS}$, should *not* be used to calculate the strains and stress resultants.

4.4.4 Complex response method

The uncoupling of the global equilibrium equations through the introduction of generalized coordinates is very efficient for a large variety of problems in structural dynamics. Occasionally, however, situations are encountered where some of the system matrices, particularly the stiffness and damping, are not entirely constant but contain some frequency dependent terms. An example is the analysis of soil–structure interaction where the soil properties are frequency dependent. Aside from selecting a predominant frequency and evaluating the matrices as if they were constant, an uncoupled solution in generalized coordinates may not be attainable.

One recourse is to employ a direct integration scheme in the time domain, as discussed in the next section. But, a more elegant approach for linear systems is to carry out the solution in the frequency domain. This is facilitated by the transformation of the global equilibrium equations, Eqs 4.11, using the very efficient Fast Fourier Transform (FFT) technique.[5] Assuming that both the stiffness and the damping matrices are composed of constant and frequency dependent parts, i.e.

$$[\bar{K}^i] = [K^i] + [\tilde{K}^i(\omega)] \tag{4.48a}$$

$$[\bar{C}^i] = [C^i] + \frac{1}{\omega}[\tilde{C}^i(\omega)] \tag{4.48b}$$

the transformed equations take the complex form

$$[[K^i] + [\tilde{K}^i(\omega_s^i)] + i\omega_s^i[C^i] + i[\tilde{C}^i(\omega_s^i)] - (\omega_s^i)^2[M^i]]\{\Delta_s^i\} = \{\mathcal{F}^i(\omega_s^i)\} \tag{4.49}$$

for a discrete frequency ω_s^i.

Note that the time dependent input vector $\{\bar{\mathcal{F}}^i(t)\}$ has been transformed, as well, into $\{\bar{\mathcal{F}}^i(\omega)\}$, which has real and imaginary parts. In the case of earthquake analysis due to uniform ground motion, Section 4.4.3, $\{\bar{\mathcal{F}}^i(t)\}$ represents the effect of the base acceleration $\ddot{z}_g^i(t)$ and the time dependent interactive forces between the foundation and the structure. The transform of the base acceleration is $\ddot{z}_g^i(\omega)$, the complex amplitude.

Equation 4.49 is solved at a sufficient number of discrete frequencies to accurately represent the response. This is dependent on the form of the input. Considering a component of the solution Δ_{si}^i, the amplitude of the response and the phase angle by which it lags the excitation are, respectively,

$$\text{Amplitude} = |\Delta_{si}^i| = [\text{Re}(\Delta_{si}^i)^2 + \text{Im}(\Delta_{si}^i)^2]^{1/2} \tag{4.50a}$$

$$\text{Phase angle} = \tan^{-1}\frac{\text{Im}\,\Delta_{si}^i}{\text{Re}\,\Delta_{si}^i} \tag{4.50b}$$

Peaks in the plots of Δ_{si}^i versus frequency indicate the damped natural frequencies. An example is shown in Fig. 4.28.

The solution for each component of $\{\Delta_i^i\}$, as a function of time, is constructed by using the inverse FFT on the corresponding complex components of $\{\Delta_{si}^j\}$. An application of this technique is presented in Section 4.6.4.5.

4.5 Direct integration solution

The solution of the $5(N+1)$ second order Eqs 4.11 may be carried out by direct integration in the time domain. This approach is an alternative to the modal and complex response methods for linear problems and a necessity for nonlinear problems. In some cases, direct integration may prove more efficient even for linear problems, particularly if time histories of displacements and stresses are to be calculated.

To accomplish this solution, the equations are first reduced to $5(N+1)$ algebraic equations by the application of unconditionally stable difference formulas. Single step, higher derivative formulas like those in the Newmark-β (with $\beta = 0.25$), or Wilson-θ schemes ($\theta = 1.4$), or multistep types such as Houbolt's method may be used. The resulting equations are then solved step-by-step for the given digitized loading function.[8]

Since time integration solutions are somewhat involved and are strongly influenced by available computer codes, details of carrying out a solution, as outlined in the preceding paragraphs, are not included here. However, there is a critical decision which is related to the properties of the shell, i.e., the selection of an appropriate time step. One wishes to use as large a step as possible, without jeopardizing the accuracy of the results. It is not easy to give precise rules for the determination of this parameter but, for the integration method described in Ref. 8 and implemented in the SHORE-III program, a maximum Δt of about 10% of the period of the highest frequency mode which is significant in the response has proven satisfactory. The reader is cautioned, however, to ensure that a chosen time step does not distort the solution.

Some examples of this solution technique are provided in Section 4.6.5 and Section 4.6.6.

4.6 Case studies

4.6.1 Free vibration of cylindrical shell

A free vibration analysis of the shell shown in Table 4.1 and in Fig. 4.2 was carried out with the SHORE-III program using seven and twelve elements and polynomials of order $n = 2, 3, 4, 5, 6$. Kinematic condensation was applied in all cases, and the results for the first four harmonics

Dynamic analysis

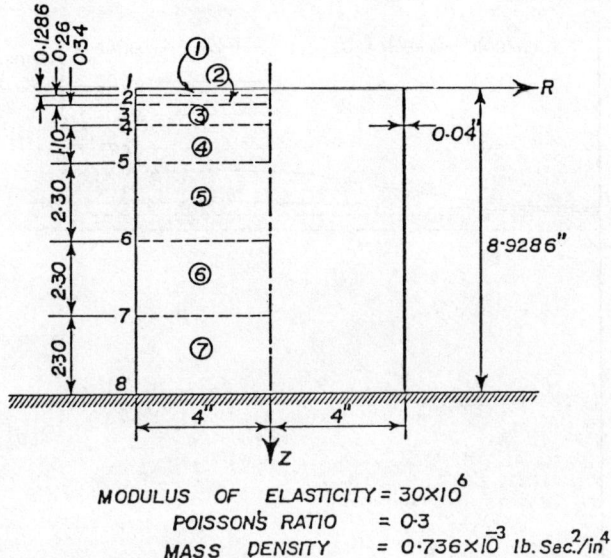

Fig. 4.2 Circular cylindrical shell

with $n = 3$ and seven elements are shown in Fig. 4.3. Comparisons with the results obtained by Donnell's theory and experiment,[9] and numerical integration[10] are also given.

The convergence characteristics with different values of n, for the case of harmonic $j = 0$, is shown in Fig. 4.4. In the first mode, the solution

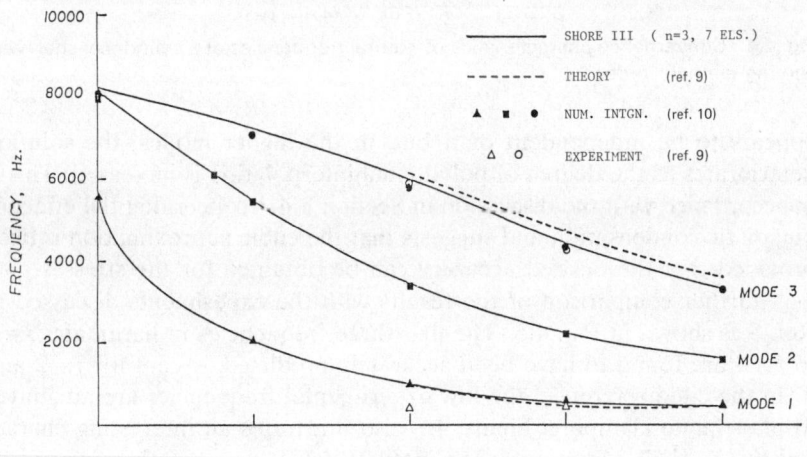

Fig. 4.3 Comparison of natural frequencies of a clamped-free cylindrical shell

91

Finite element analysis of shells of revolution

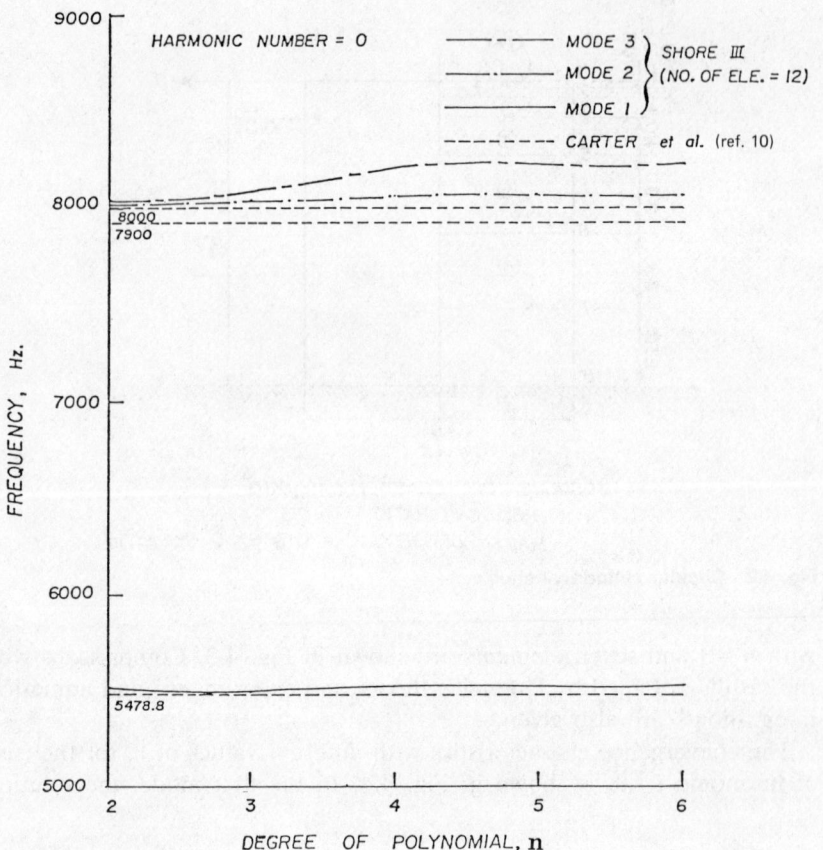

Fig. 4.4 Convergence characteristics of natural frequencies of a cylindrical shell with varying element order

appears to be independent of n but, in the higher modes, the solution deteriorates as the degree of polynomial interpolation is increased. This is in accordance with the discussion in Section 4.4.1 concerning the effect of kinematic condensation and suggests that the cubic approximation is best, provided that the desired accuracy can be obtained for the stresses.

A further comparison of the results with the experiments discussed in Ref. 9 is shown in Fig. 4.5. The first three frequencies in harmonics $j = 0$ to $j = 8$ are found to have been accurately predicted, except for $j = 2$ and 3. In the cited reference, the low experimental frequencies are attributed to inadequate clamping. Figure 4.5 also illustrates an interesting characteristic of shell vibrations. The natural frequency for the first mode *decreases* as the harmonic number j increases until a minimum is reached, whereupon the frequency *increases* with j. The same characteristic is

Dynamic analysis

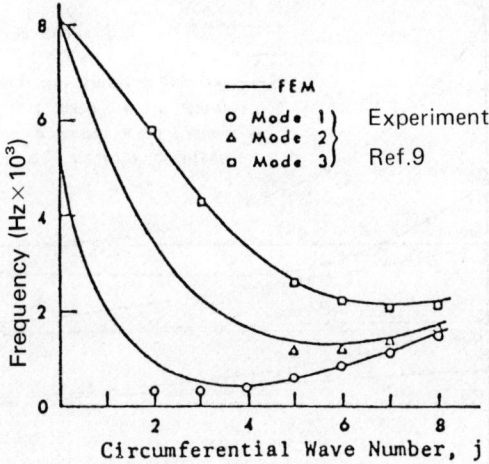

Fig. 4.5 Fixed-free cylindrical shell

present in the buckling of similar shells and has been vividly demonstrated in experiments, where the lowest vibration or buckling mode contains several circumferential waves.

4.6.2 Free vibration of hemispherical shell

The hemispherical shell shown in Fig. 4.6 is discretized with three, five and ten equally spaced elements and analyzed for the $j = 0$ harmonic

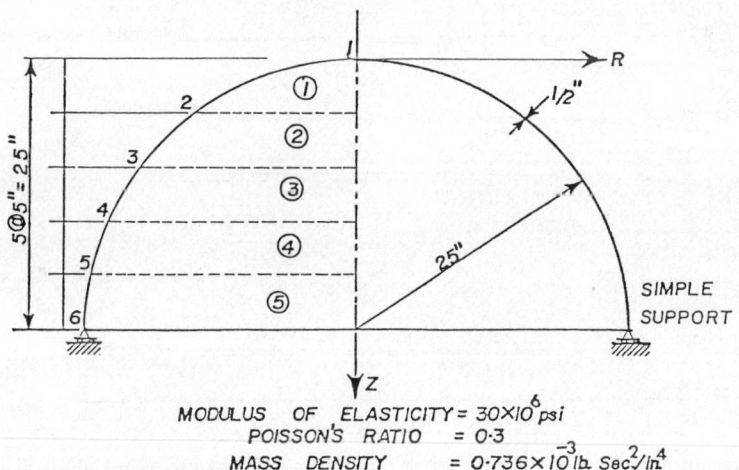

Fig. 4.6 Hemispherical shell

Finite element analysis of shells of revolution

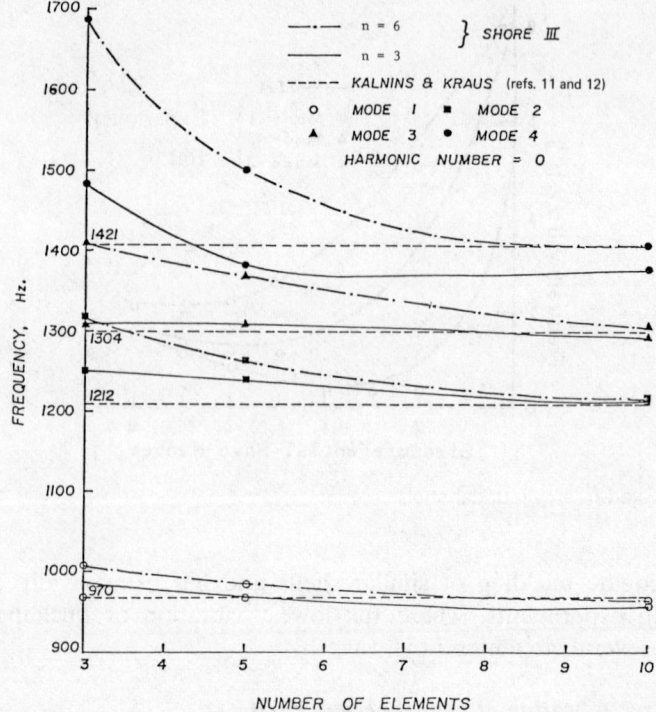

Fig. 4.7 Natural frequencies of a simply supported hemispherical shell

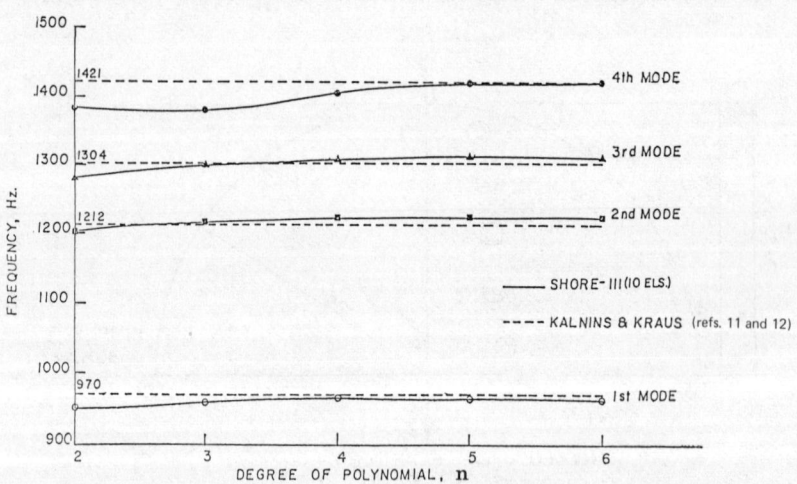

Fig. 4.8 Convergence characteristics of natural frequencies of a hemispherical shell with varying element order

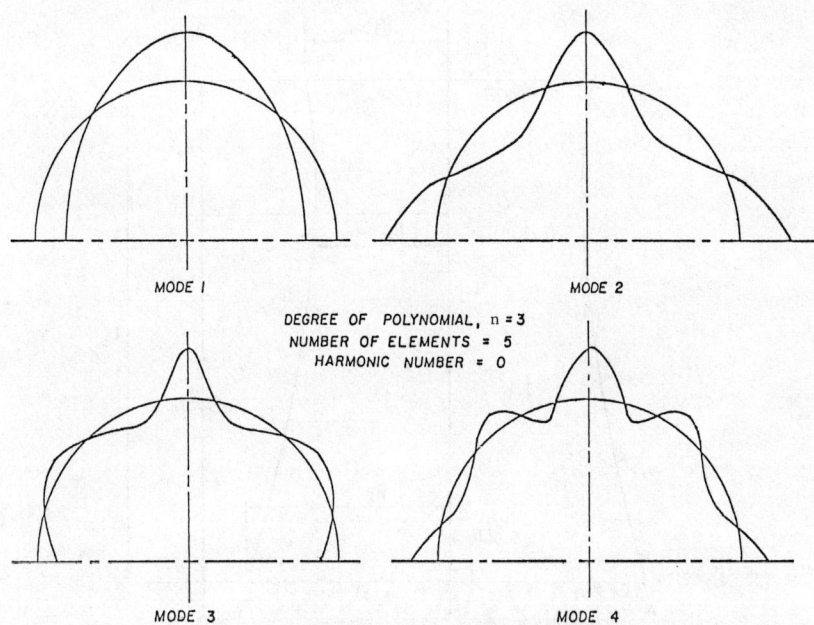

Fig. 4.9 Mode shapes of a hemispherical shell using SHORE-III

using third and sixth order polynomial interpolations. The natural frequencies for the first four modes are shown in Fig. 4.7 along with results from an analytical solution obtained by Kalnins and Kraus[11,12] based on Reissner–Naghdi theory, which includes both transverse shear and rotary inertia effects.

The previously indicated superiority of the third order approximation, as opposed to the sixth order, is maintained, except in the fourth mode where the $n = 6$ solution is superior. This is demonstrated more clearly by Fig. 4.8, which indicates that the accuracy of the lower order condensed solutions tends to deteriorate for high modes. From a practical standpoint, it is probably advisable to perform a convergence check by using two or more values of n; by using discretization refinement; or both.

The mode shapes for the normal displacement of the hemispherical shell are shown in Fig. 4.9.

4.6.3 *Dynamic analysis of column-supported cooling tower shell*

4.6.3.1 Description of shell

A cooling tower shell supported on a system of inclined columns as shown in Fig. 4.10 is considered. The dimensions chosen are representative of modern large towers, except that the thickness is usually varied

Finite element analysis of shells of revolution

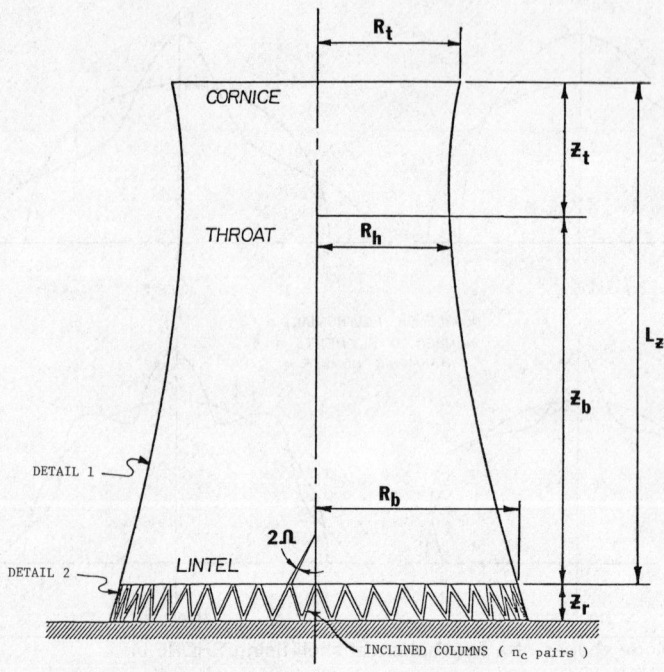

Fig. 4.10 Hyperbolic cooling tower

along the meridian in actual shells. Here, an average thickness is used. The results of this analysis have been presented in detail elsewhere.[6,13]

4.6.3.2 Free vibration analysis

The natural frequencies for four base conditions are given in Table 4.2. The influence of the flexible base, which represents the supporting columns, is to reduce the frequencies substantially. These results were based on a model which did not represent the column mass in a consistent fashion. A modest difference was obtained when a consistent mass matrix, as derived in Section 4.3.2, was used, as indicated by the parenthetical values in Table 4.2.[3]

The corresponding mode shapes are given in Fig. 4.11. The predominant influence of the flexible base is on the circumferential (v) and normal (w) displacements.

4.6.3.3 Response spectrum analysis

A horizontal earthquake was represented by a design spectrum based on a 0.12 g ground acceleration and 4% of critical damping. The velocity

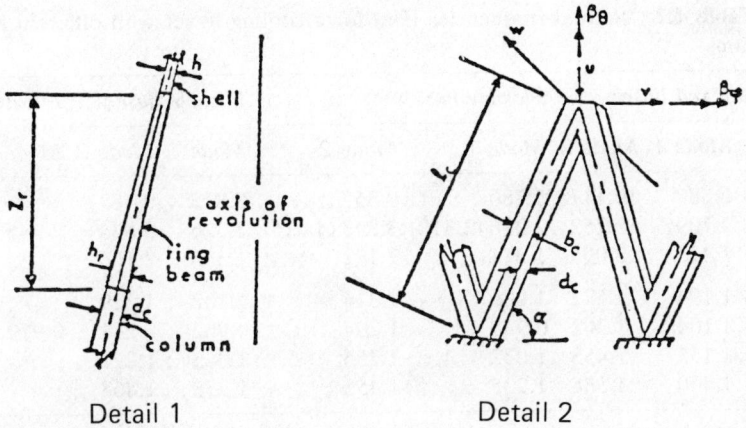

Detail 1   Detail 2

Basic dimensions

$Z_b = 350\cdot 86$ ft, $Z_t = 98\cdot 89$ ft, $Z_r = 32\cdot 0$ ft

$R_b = 182\cdot 25$ ft, $R_h = 116\cdot 5$ ft, $R_t = 123\cdot 0$ ft

$h = 12\cdot 0$ in., $h_r = 24\cdot 0$ in., $l_c = 41\cdot 0$ ft, $b_c = 52\cdot 0$ in.

$d_c = 24\cdot 0$ in., $n_c = 44$, $\alpha = 71°$

$E_{\text{shell}} = E_{\text{column}} = 4 \times 10^6$ psi

$\mu_{\text{shell}} = \mu_{\text{column}} = \frac{1}{6}$, mass density $= 0\cdot 225 \times 10^{-3}$ lb-s^2/in^4

Thickened ring beam and deeper columns

$h_r = 42\cdot 0$ in., $b_c = 30\cdot 0$ in., $d_c = 42\cdot 0$ in.

(All other dimensions same as above.)

Fig. 4.10 (contd.)

spectrum ordinates for the first three modes of the $j = 1$ harmonic are given in Table 4.3 for the 'fixed base' and 'flexible base without ring beam' cases of Table 4.2. Also given in the table are the maximum values of the generalized coordinates η_r^i, as computed from Eqs 4.44 and 4.45.

From the mode shapes and generalized coordinates, the displacements u_r^1, v_r^1 and w_r^1 have been calculated.[6] Here, the RSS combined displacements, Eq. 4.47, as well as the ABS combination, are shown on Figs 4.12–4.14. The pronounced effect of the column flexibility is

Finite element analysis of shells of revolution

Table 4.2 Natural frequencies (Hz) for a cooling tower with different base conditions

j	Fixed base*		Flexible base†		Flexible base‡		Flexible base§	
	Mode 1	Mode 2	Mode 1	Mode 2	Mode 1	Mode 2	Mode 1	Mode 2
0	6.588	10.117	5.780	9.557	5.873	9.487		
1	2.709	5.752	2.296 (2.318)	3.893 (4.015)	2.326	3.834	2.345	3.749
2	1.475	3.095	1.311	2.172	1.333	2.166		
3	1.194	1.672	1.086	1.314	1.102	1.329		
4	1.104	1.302	0.945	1.204	0.962	1.211	0.979	1.240
5	1.131	1.453	1.032	1.256	1.045	1.272		
6	1.400	1.568	1.235	1.455	1.258	1.468		

* Without the ring beam.
† Without the ring beam.
‡ With the ring beam.
§ With increased dimensions of the ring beams.
( ) Consistent mass used for columns.

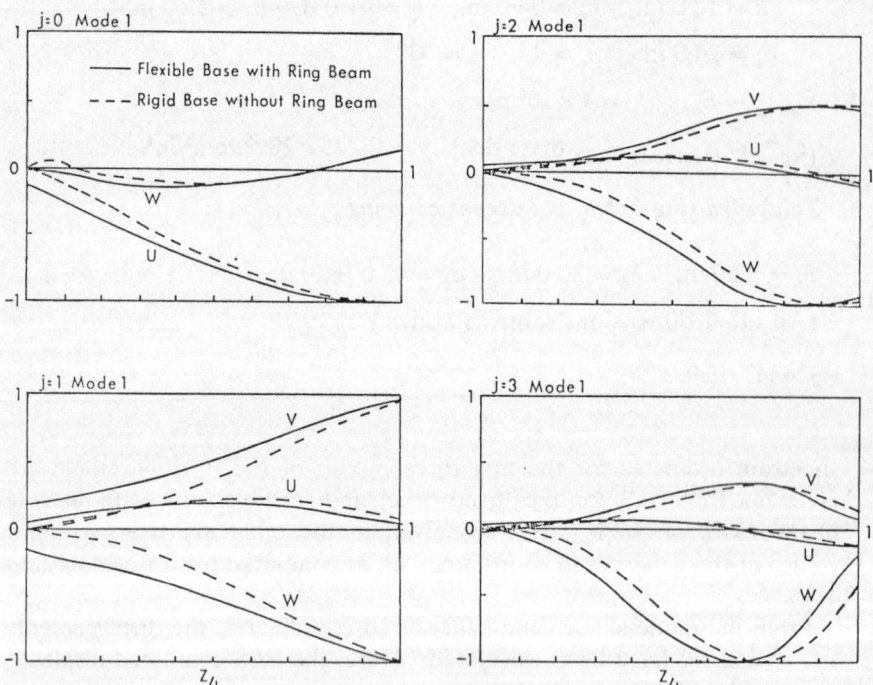

Fig. 4.11 Mode shapes

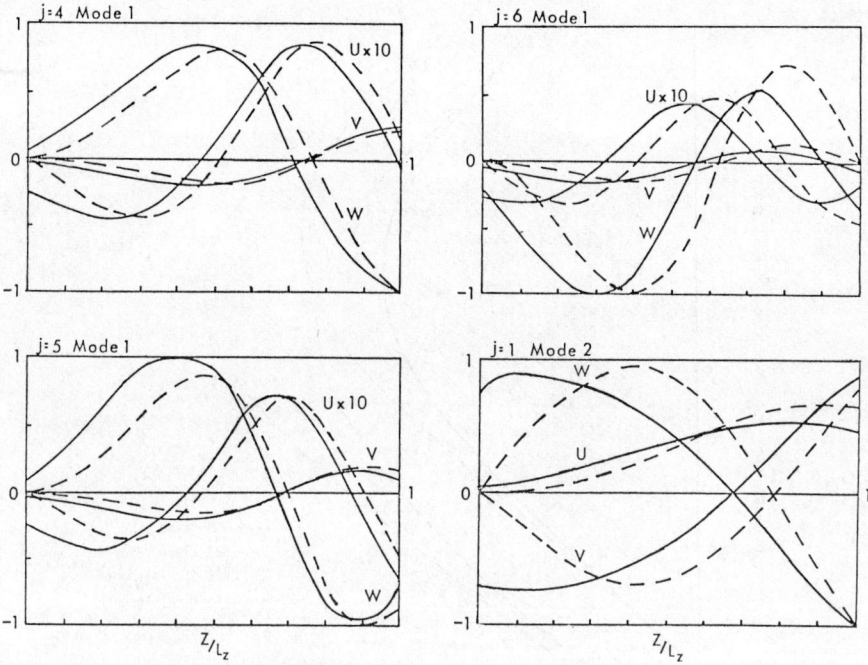

Fig. 4.11 (contd.)

obvious. It is also apparent that the ABS combination gives considerably larger values at most locations.

Finally, the stress resultants in the shell may be computed for the response spectrum loading case. The details of the calculations are developed elsewhere.[13] For the purposes of this illustration, it is only necessary to mention that the preferred method of computing the stresses is to apply the kinematic and constitutive laws, Eqs 2.29 and 2.45 respectively, to the displacement vector for each mode, Eq. 4.46, and then to sum the modal contributions. Values of N_ϕ and N_θ are shown in Figs 4.15 and 4.16 and again, the ABS combination may be conservative.

Table 4.3

r mode	Fixed base			Flexible base		
	$\omega_r^1/2\pi$, Hz	S_{Vr}^1 in/s	$(\eta_r^1)_{max}$	$\omega_r^1/2\pi$, Hz	S_{Vr}^1 in/s	$(\eta_r^1)_{max}$
1	2.7087	4.629	0.4207	2.2966	5.092	0.5488
2	5.7513	2.432	0.0532	3.8892	3.609	0.0921
3	9.1177	1.237	0.0155	7.7279	1.528	0.0013

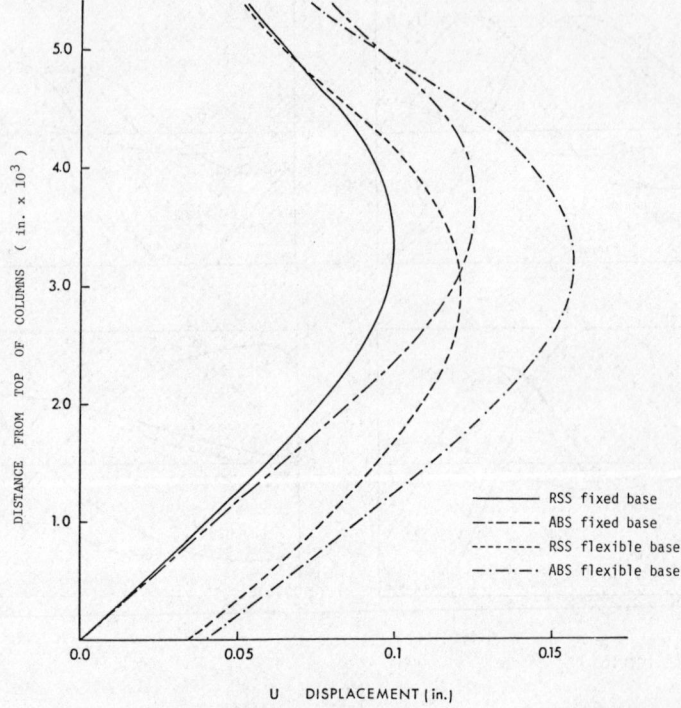

Fig. 4.12 Maximum meridional displacement at $\theta = 0°$ due to earthquake

Note that the static correction for the discrete column supports (Section 3.3.6) was *not* applied in this example, but a more detailed stress analysis including this effect is provided in the next example.

4.6.4 *Cooling tower on an interactive foundation*

4.6.4.1 Shell model

The plots of the displacements for the shell in the previous example, shown in Figs 4.12–4.14, indicate that consideration of the column flexibility leads to significant relative displacements between the ends of the columns. Correspondingly large forces and moments would be introduced into these critical members. However, the response of the columns, and of the shell itself, would be altered if the lower ends of the columns were not completely fixed, but deformed with the supporting medium. This is known as soil–structure interaction and has been studied extensively, particularly in regard to the response of structures to seismic loading.[14]

Although a formulation for an interactive soil medium cannot be fully

Dynamic analysis

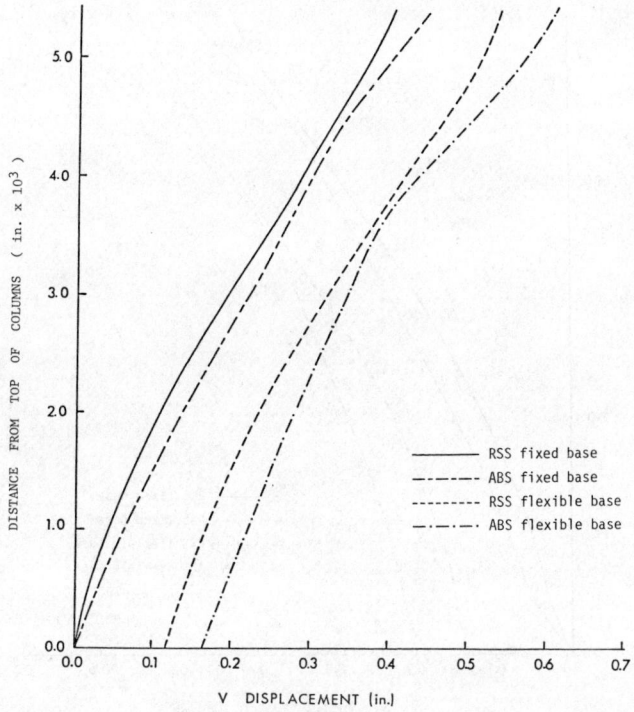

Fig. 4.13 Maximum circumferential displacement at $\theta = 90°$ due to earthquake

presented within the scope of this book, it is informative to assess some results. We may account for this effect by adding two components to the cooling tower model, a ring footing and a set of linear springs, to depict the soil reactions, as shown in Fig. 4.17,[15] along with the high precision rotational shell elements and the open element previously derived. The lintel and column regions are expected to be influenced by the static correction discussed in Section 3.3.6, applied by superposition to the continuous boundary solution.

We focus on the representation of the soil by a dynamic boundary at the common degrees of freedom between the shell foundation and the underlying soil. The finite element model of the soil medium, along with the Equivalent Boundary System (EBS) at the footing base, are shown in Fig. 4.18. The EBS is formulated to represent a soil medium which may be layered strata underlain by rock or an elastic half-space. The EBS is *frequency dependent* and should ideally be updated for each Fourier harmonic in the circumferential direction (θ coordinate). The finite element model used in formulating the EBS, Fig. 4.18, consists of axisymmetric, isoparametric quadratic solid elements for the core region, which

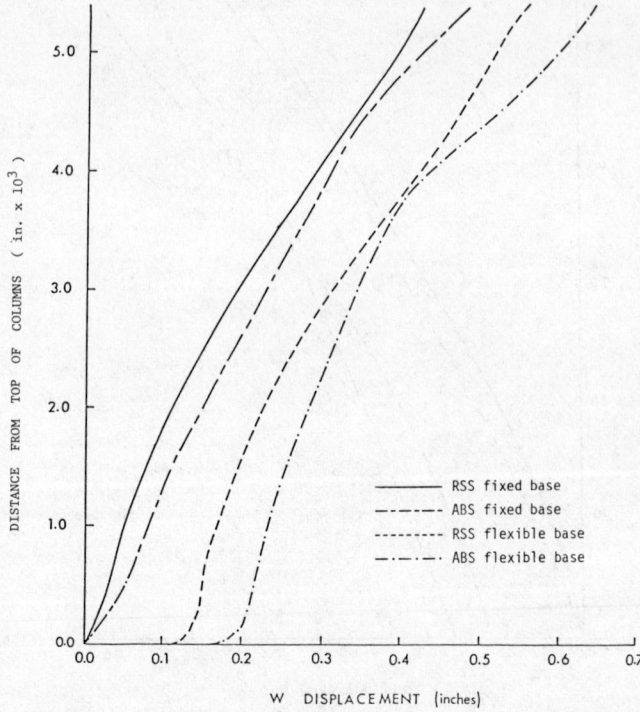

Fig. 4.14 Maximum normal displacement at $\theta = 0°$ due to earthquake

is bounded vertically at the foundation extremities. The far field is represented as a semi-analytic, energy transmitting boundary, which is based on the exact displacement functions in the horizontal direction and an expansion in the vertical direction consistent with that used for the core elements.[16]

Eventually, the properties of the soil medium properties are reduced to those of linear springs for each of the shell DOF, as shown on Fig. 4.17 and on Fig. 4.18(b). The lower boundary must be fixed to ensure the numerical stability of the finite element solution. Parametric studies have indicated that in the case of no actual rock bed (very deep strata), the assumption of a fixed lower boundary at a depth ratio $H/R_0 = 3$ is adequate.[17] Additional details of the model of the soil medium are also given in Ref. 17.

The computations are carried out using finite element techniques which are similar to those developed in the earlier chapters. Special features of the analysis include a check for uplift, depicted in Fig. 4.19, and the static correction for the effect of the discrete columns on the shell, as shown in Fig. 3.11.

Dynamic analysis

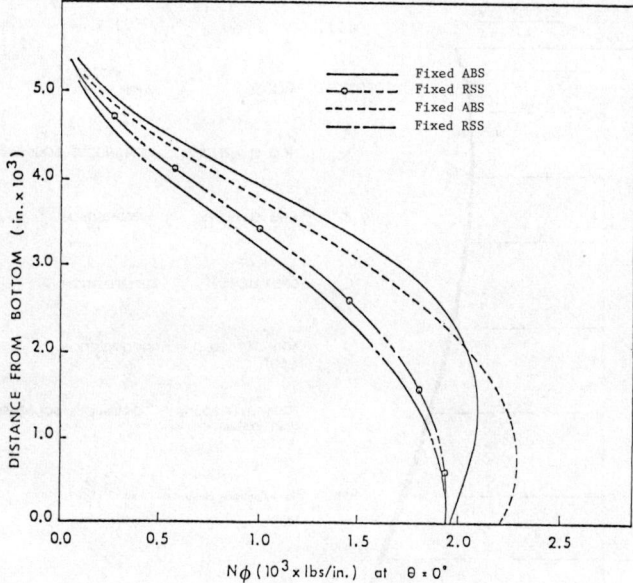

Fig. 4.15 N_ϕ response spectrum analysis for different continuous end conditions

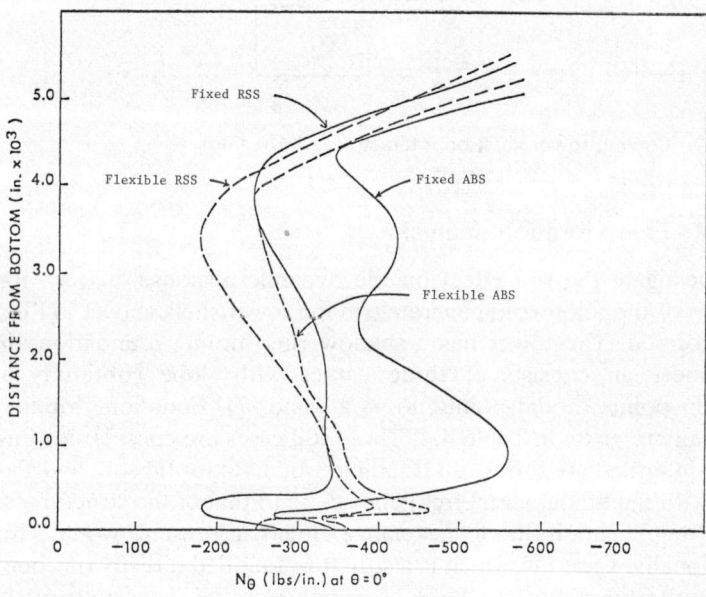

Fig. 4.16 N_θ response spectrum analysis for different continuous end conditions

Finite element analysis of shells of revolution

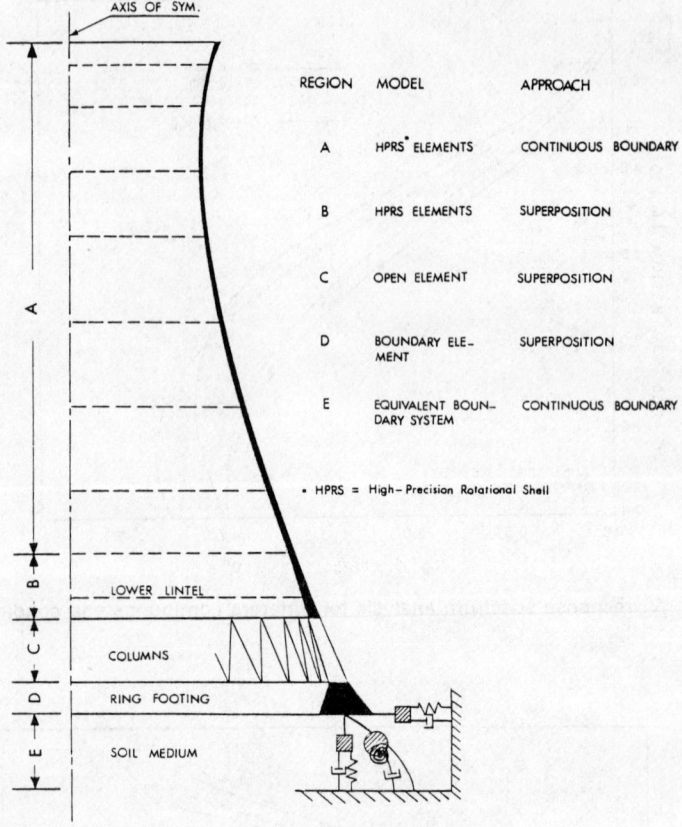

Fig. 4.17 Cooling tower shell on interactive soil medium

4.6.4.2 Free vibration analysis

To investigate the soil effect on the dynamic response, a free vibration analysis of the reinforced concrete cooling tower shell, shown in Fig. 4.20, is performed. The tower has a shallow ring footing foundation and the shell meridian consists of three curves with slope continuity at the junction points (nodal points no. 4 and no. 7). Equations for the shell meridian are given in Table 4.4. Three soil cases are considered: Case I, a soft to intermediate soil; Case II, an intermediate to stiff soil; and Case III, a soil with the fundamental frequency close to that of the structure, so that strong amplification due to resonance effects, if present, would show up. Additionally, Case IV, which is a structure founded directly on competent rock, is considered as the basis of comparison.[18] (The details of the soil properties, given in Ref. 17, are summarized in the Appendix.)

Dynamic analysis

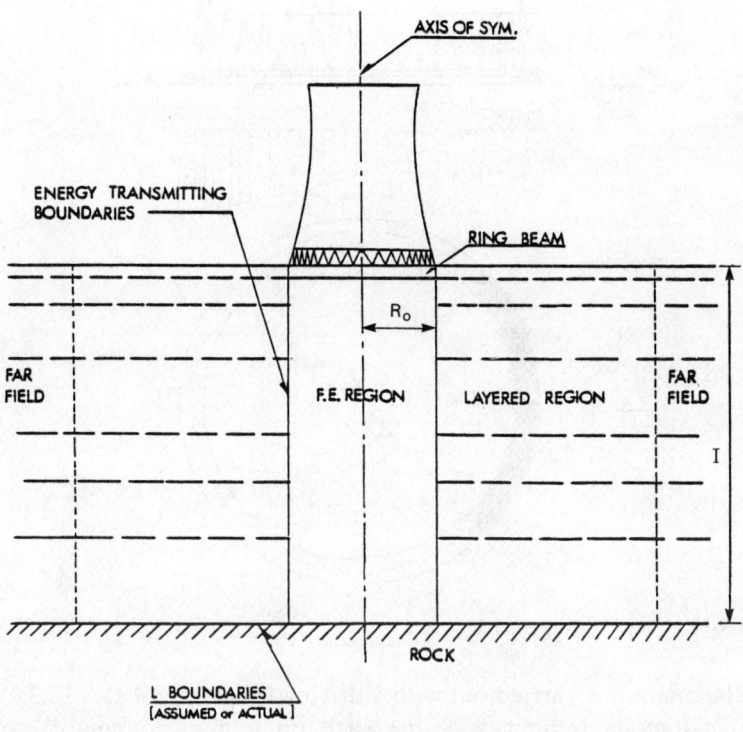

Fig. 4.18(a) Finite element model for the soil medium

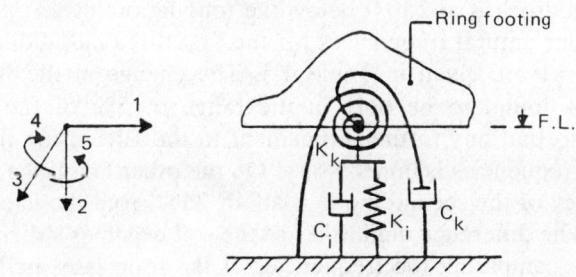

i = 1,2,3 (Translational DOF)
k = 4,5 (Rotational DOF)

Fig. 4.18(b) Equivalent boundary system

Finite element analysis of shells of revolution

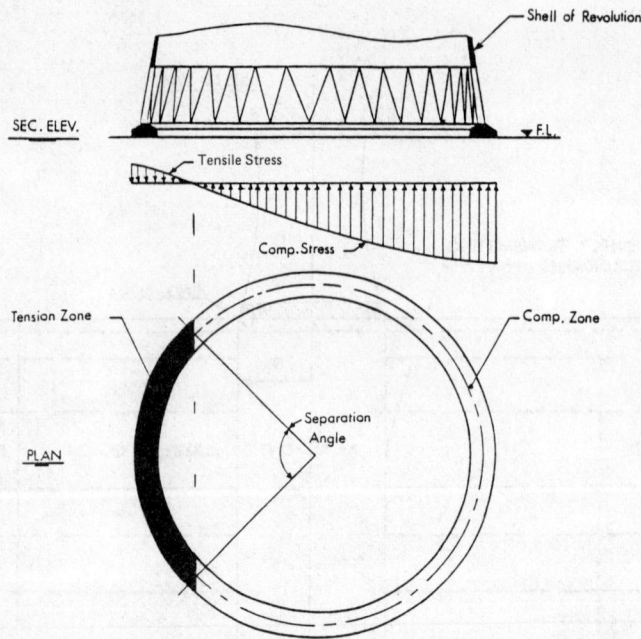

Fig. 4.19 Base uplift

The analysis is carried out with a driving frequency of $\Omega = 12.34$ rad/s, the first mode frequency of the shell on a *fixed* foundation for the antisymmetrical mode, $j = 1$. The EBS consists of a soil model with 24 elements and a depth ratio, H/R_0, in the range of 3, except for Case II where the bedrock is at 250 ft below the foundation level.

The circular natural frequencies for the first three meridional modes in harmonic $j = 1$ are given in Table 4.5. The change in the fundamental frequency is found to be only in the range of 5% of the fixed base frequency, so that any further refinement in the EBS using the resulting interactive frequencies is unnecessary. On the other hand, the decrease in the frequency of the second mode is about 25% from the fixed base case (Case IV). The difference diminishes as the soil becomes stiffer, as may be observed by comparing the frequencies of the four cases in Table 4.5.

In Fig. 4.21, the first three normalized eigenvectors for Cases I and IV, which represent the extreme soil conditions, are shown. For the soft to intermediate soil case, Case I, the interactive eigenvectors of the second mode are drastically different than the fixed case, Case IV, whereas there is not much difference between the eigenvectors of the first mode for the two soil cases. A similar but less predominant influence of the soil on the interactive eigenvectors is observed for Case II (the stiff-shallow soil

Dynamic analysis

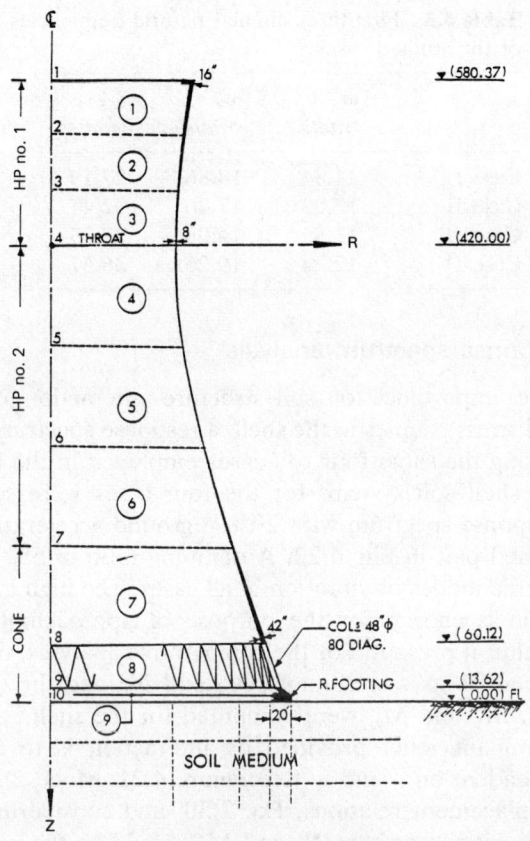

Fig. 4.20 Cooling tower on a hypothetical foundation

case). The eigenvectors of Case III were found to be very similar to those of Case I. This may be attributed to the combined effect of the soil depth and the shear modulus, producing very similar compliances for the first and the third soil cases. The change in the second meridional mode for the more flexible soil may be expected to affect the stress distribution in the lower region of the shell, which will be confirmed in the next section.

Table 4.4 Shell meridian of the structure under study

Shell type	Nodes From	To	Equation
HP no. 1	1	4	$Z^2 - 123.68377\ R^2 + 27587.5165\ R - 1536846.5 = 0$
HP no. 2	4	7	$Z^2 - 9.40153\ R^2 + 1302.5923\ R - 25462.9 = 0$
Cone	7	10	$0.3Z - R + 87.2112 = 0$

107

Table 4.5 First three circular natural frequencies of the studied cases

	ω_1^1 (rad/s)	ω_2^1 (rad/s)	ω_3^1 (rad/s)
Case I	11.84	14.86	27.19
Case II	12.02	17.46	32.41
Case III	11.85	14.98	27.33
Case IV	12.34	19.26	36.37

4.6.4.3 Response spectrum analysis

To assess the importance of soil–structure interaction on the stress resultants and stress couples in the shell, a response spectrum analysis was carried out using the same four soil cases employed in the free vibration analysis. The shell–soil systems for the four cases were subjected to a horizontal response spectrum with 20% g ground acceleration, as shown on the combined plot in Fig. 4.22. A damping ratio of 5% is considered for the first three modes of vibration in all cases. The high intensity of the ground motion is chosen for the purpose of approaching the case of foundation uplift, if present. For the comparative analyses, the meridional and circumferential stress resultants, N_ϕ and N_θ, and the corresponding stress couples, M_ϕ and M_θ, were computed for the shell.[15]

The predominant relief provided by interaction is to reduce forces primarily dependent on v and w (see Figure 4.21, Mode 2). Referring to the strain–displacement relations, Eq. 2.30, and considering β_ϕ as proportional to $w_{,s}$, it is seen that N_θ and M_ϕ should be the quantities most affected and this was confirmed by the results.[15] The reduction in seismically induced forces, which is generally in the range of 20% from the fixed base solution, may permit a corresponding reduction of the shell cross-sectional thickness and the horizontal reinforcing steel.

The axial forces, bending moments, and twisting moments in the columns were calculated at $\theta = 0°$ and the results are shown in Table 4.6. It may be observed that there is a *decrease* in the axial forces and bending moments, more than 50% for some soil conditions, as the soil stiffness *decreases*. The decrease in the bending moment is attributed to the smoothing of the second mode shape, Fig. 4.21, in the lower region, while the reduction of the axial forces from the fixed base case occurs despite a considerably greater second mode contribution. It is apparent that any valid analysis for this type of problem must include at least two meridional modes.

The twisting moment in the columns increases as the soil stiffness decreases. However, the values of the twisting moments are not large

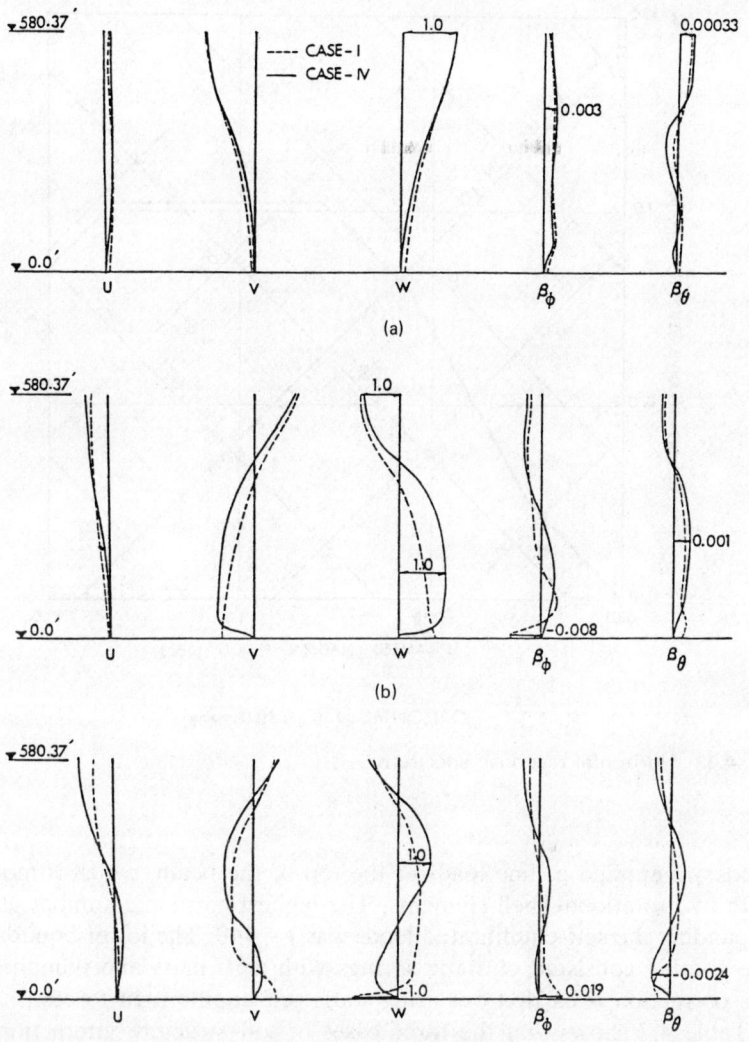

Fig. 4.21 Soil effect on the antisymmetrical eigenvectors of a cooling tower. (a) Mode no. 1; (b) mode no. 2; (c) mode no. 3

enough to be a controlling factor in the column design, as may be seen from Table 4.6. The response of the concentric ring footing is given in Table 4.7, both directly beneath the column and between the columns (Field). The results presented in this table are the complete solutions, which consist of the continuous boundary solution and the self-equilibrated correction. In the correction, the resulting self-equilibrated

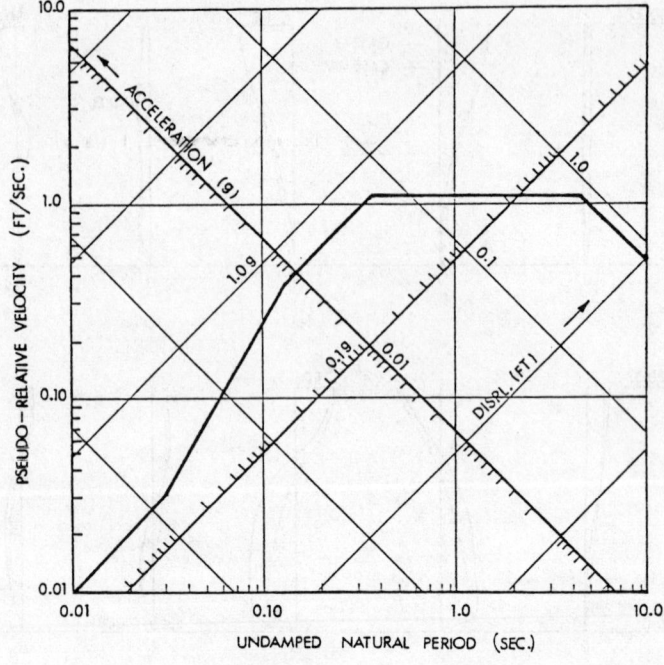

HORIZONTAL [0.20 g , 5% Damping]

Fig. 4.22 Horizontal response spectrum

loads are applied as line loads at the top of the beam, which is modeled with two rotational shell elements. The highest harmonic number used in expanding the self-equilibrated loads was $\bar{j} = 440$. The lower boundary of the footing consisted of static springs with zero mass and damping, i.e. the correction is carried out using static self-equilibrated forces.

Table 4.7 shows that the three cases of soil–structure interaction give responses for the axial forces and bending moments in the ring footing which are sharply *higher* than the fixed base response, while the torsional moment decreases as the soil becomes softer. Incidentally, the values presented for the vertical moment are computed from the N_θ distribution along the footing depth, since the vertical bending moment corresponds to the rotational degree of freedom about the normal axis, which is neglected in shell theories.[19]

To check against foundation uplift, the N_ϕ component of the stress resultants is computed at the foundation level for dead load (DL) which is factored by 0.9 and then added to the unfactored earthquake response.

Dynamic analysis

Table 4.6 Maximum column forces at $\theta = 0°$ (horizontal ground motion)

	Axial force (kips)				Bending moment (ft-kips)				Twisting moment (ft-kips)			
Mode	Case I	Case II	Case III	Case IV	Case I	Case II	Case III	Case IV	Case I	Case II	Case III	Case IV
1	539.7	592.9	543.3	826.8	160.2	88.8	153.6	160.5	−14.4	−1.7	−13.1	−6.1
2	−450.1	−364.4	−451.2	−141.3	166.5	143.4	157.9	615.1	−30.3	7.8	−28.6	−17.2
3	0.2	−42.2	1.1	−19.6	1.0	398.8	9.4	17.5	−0.1	16.4	−0.3	−0.1
RSS	±702.2	±697.2	±706.2	±838.9	±240.3	±432.9	±220.5	±636.0	±33.7	±18.2	±31.5	±18.2

Table 4.7 Foundation response to horizontal ground motion (RSS)

Soil case	θ	Axial force (kips)		Vertical moment (ft-kips)		Horizontal moment (ft-kips)		Torsion (ft-kips)	
		Column	Field	Column	Field	Column	Field	Column	Field
Case I	0°	±1867.4	±1805.7	∓10953.2	∓11837.4	∓809.2	∓911.6	±3767.7	±4340.5
	90°	±152.4	±157.9	∓715.7	∓731.6	∓32.7	∓64.4	±7762.0	±7699.3
Case II	0°	±1385.0	±1361.6	∓6705.0	∓7007.8	∓620.7	∓701.0	±5013.1	±4897.6
	90°	±141.6	±150.7	∓682.3	∓695.9	∓24.1	∓54.9	±9267.4	±9226.3
Case III	0°	±1751.1	±1691.8	∓9817.4	∓10911.1	∓752.5	∓839.6	±4266.1	±4703.8
	90°	±146.7	±152.2	∓700.5	∓715.9	∓30.8	∓62.8	±8117.7	±8100.4
Case IV	0°	±392.4	±388.8	∓745.7	∓699.0	∓110.0	∓118.1	±6946.2	±5988.1
	90°	±23.0	±27.7	∓47.7	∓50.6	∓6.7	∓7.2	±11207.0	±11200.6

Table 4.8 N_ϕ-component at F.L. ($j = 1$) (units: k/ft)

DL	0.9(DL)	Case I		Case II		Case III		Case IV	
		EQ	Net	EQ	Net	EQ	Net	EQ	net
−77.8	−70.0	64.8	−5.2	59.2	−10.8	61.9	−8.1	58.3	−11.7

The results are tabulated in Table 4.8. It may be seen that the net stress at the foundation level for all cases is compressive, and that no uplift occurs for the relatively severe 20% g spectrum used in the analysis. However, it is clear that the softer the soil, the more likely it is for uplift to occur. To investigate this possibility more closely, the vertical component of the earthquake may be included. A vertical response spectrum with 13% g ground acceleration and 5% damping was added, and an analysis carried out at a driving frequency of $\Omega = 32.75$ rad/s, which is the fundamental frequency of the structure on a *fixed* base for $j = 0$. The RSS value of N_ϕ at the foundation level for the vertical and horizontal ground motions is computed, and the net value of N_ϕ is found by combining the resulting RSS value of N_ϕ with the factored DL value:

$$N_\phi^2(\text{net}) = (N_{\phi X}^2 + N_{\phi Z}^2)^{1/2} - 0.9 N_{\phi D}^2 \tag{4.51}$$

in which $N_{\phi X} = N_\phi$ at the foundation level due to horizontal ground motion; $N_{\phi Z} = N_\phi$ at the foundation level due to vertical ground motion; and $N_{\phi D} = N_\phi$ at the foundation level due to the dead load.

For Case I, N_ϕ (net) is calculated from Eq. 4.51 with $N_{\phi Z} = 29.8$ kips/ft and the resulting value of N_ϕ (net) is found to be a tension of 1.3 kips/ft, which indicates uplift. However, N_ϕ (net) is probably too small to cause actual uplift as this net stress could be counteracted by the soil friction on the sides of the footing.

4.6.4.4 Translation and rocking

For the softer soils, it may be important to consider translation and rocking of the foundation itself with respect to the ground. For tall frames, a representative model has been formulated by Shye and Robinson[20] and this concept has been extended to the rotational shell configuration by Lee and Gould.[21,22]

As shown in Fig. 4.23(a), a uniform vertical translation u_b is introduced for the $j = 0$ harmonic, and a uniform horizontal translation w_b and rocking angle $\beta_{\phi b}$ are indicated for the $j = 1$ harmonic. Before these terms can be added to $\{D_i^i(0)\}$, the displacement vector at node $s = 0$ on element i as defined in Eq. 3.6, they must be transformed into the curvilinear

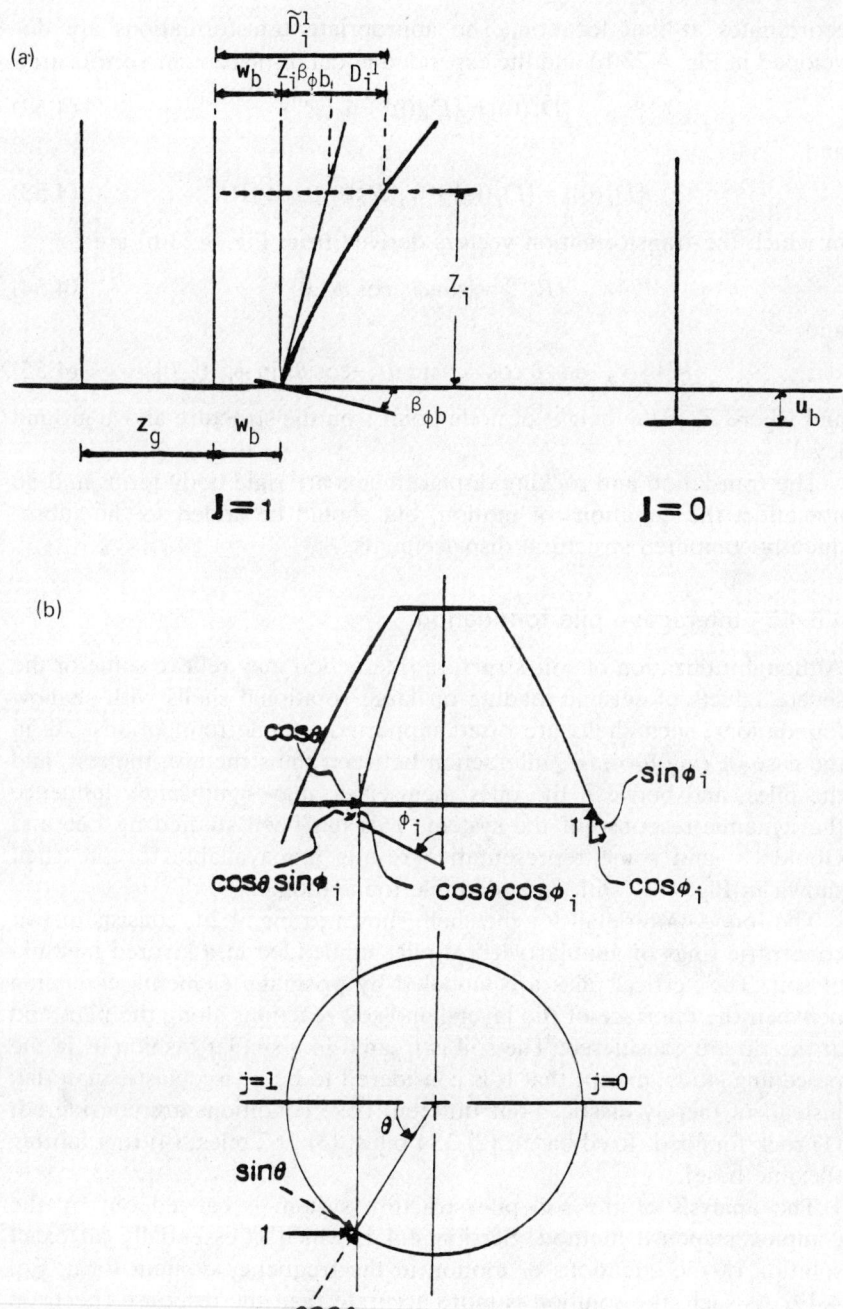

Fig. 4.23 (a) Rigid-body displacements (b) Transformation vector $\{R_i^j\}$

coordinates at that location. The appropriate transformations are developed in Fig. 4.23(b) and the expanded nodal displacement vectors are

$$\{\tilde{D}_i^0(0)\} = \{D_i^1(0)\} + u_b\{R_i^0\} \tag{4.52}$$

and

$$\{\tilde{D}_i^1(0)\} = \{D_i^1(0)\} + w_b\{R_i^1\} + \beta_{\phi b}Z_i\{R_i^1\} \tag{4.53}$$

in which the transformation vectors derived from Fig. 4.23(b) are

$$\{R_i^{(0)}\} = \{\sin \phi_i \quad \cos \phi_i \quad 0\} \tag{4.54}$$

and

$$\{R_i^{(1)}\} = \{-\cos \theta \cos \phi_i \quad \sin \theta \quad -\cos \theta \sin \phi_i \quad 0 \quad 0\} \tag{4.55}$$

and where Z_i is the height of node point i on the structure above ground level.

The translation and rocking displacements are rigid body terms and do not affect the equations of motion, but should be added to the subsequently computed structural displacements.

4.6.4.5 Interactive pile foundation

Although utilization of soil–structure interaction may relieve some of the severe effects of seismic loading on large rotational shells with shallow foundations, such shells are often supported on pile foundations. As in the case of ring footings, interaction between the structure, the soil, and the piles, and between the piles themselves, may significantly influence the dynamic response of the system. This has been studied by Lee and Gould[21,22] and some representative results are available for the shell shown in Fig. 4.20 with an added pile foundation.

The foundation detail for the shell, shown in Fig. 4.24, consists of two concentric rings of annular vertical piles embedded in a layered medium of soil. The vertical piles are modeled by prismatic elements extending between the interface of the layers, and soil reactions along the piles and at the tip are considered. The soil is treated in a similar fashion as in the preceding study, except that it is considered to be a viscoelastic material, instead of merely elastic. Four different base conditions are considered: (1) rock founded (fixed base); (2) 334 piles; (3) 167 piles; (4) ring footing (flexible base).

The analysis of the soil–pile–structure system is carried out by the complex response method, Section 4.4.4, which is essentially an exact solution of the equations of motion in the frequency domain form, Eq. 4.49. As such, the solution is more accurate than the response spectrum solution presented in the preceding study of soil–structure interaction in that the time phase relationships are maintained and the frequency

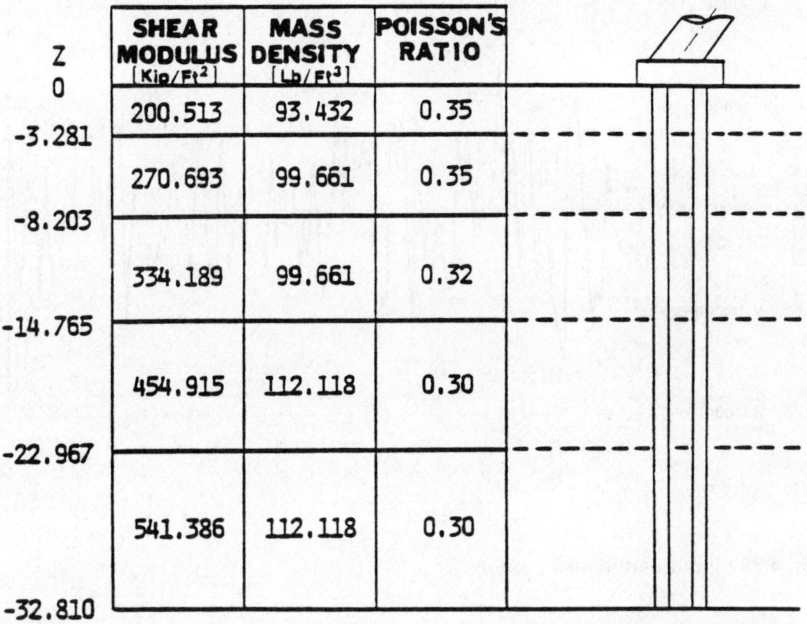

Z	SHEAR MODULUS [Kip/Ft²]	MASS DENSITY [Lb/Ft³]	POISSON'S RATIO
0			
	200.513	93.432	0.35
-3.281			
	270.693	99.661	0.35
-8.203			
	334.189	99.661	0.32
-14.765			
	454.915	112.118	0.30
-22.967			
	541.386	112.118	0.30
-32.810			

Fig. 4.24 Foundation detail for pile-supported cooling tower. Pile properties: density = 0.48 KCF; radius = 0.75 ft; thickness = 0.25 ft; length = 32.81 ft. $E_p = 4.2336 \times 10^6$ KSF; width of ring footing = 14 ft; radius of foundation = 213.2 ft; C-C distance between two piles = 8 ft (in both radial and circumferential directions); no. of piles = 334; total pile segment number = 10; vertical load on each pile = 78 Kips; hysteretic damping of soil = 10%

dependence of the foundation stiffness and damping is properly included. As an alternative, a direct integration solution, Section 4.5, can be used, and some calibration studies for a fixed base tower were carried out in Ref. 21. Very good agreement between the two solutions was achieved, but the frequency dependence of the foundation system makes the complex response method the preferable choice for the case at hand.

The tower of Fig. 4.20 with the four specified base conditions is analyzed for the El Centro earthquake (18 May 1940) EW Comp. with input in the form of Fig. 4.25. For the complex response analysis, the time step is taken as 0.02 s which gives a total of $2^9 = 512$ points for the duration of 10.22 s. The damping coefficients for the structure, as defined in Eq. 4.12, are $\alpha_1 = 0.715779$ and $\alpha_2 = 0.003356$, based on a β_r^i 5% damping ratio for the first two modes of vibration (the modes of vibration are obtained from a free vibration analysis with the base fixed).

The variation of discrete values of displacement amplitude $|w|$ at the first fundamental frequency (i.e. the first vibration mode) along the structure

Finite element analysis of shells of revolution

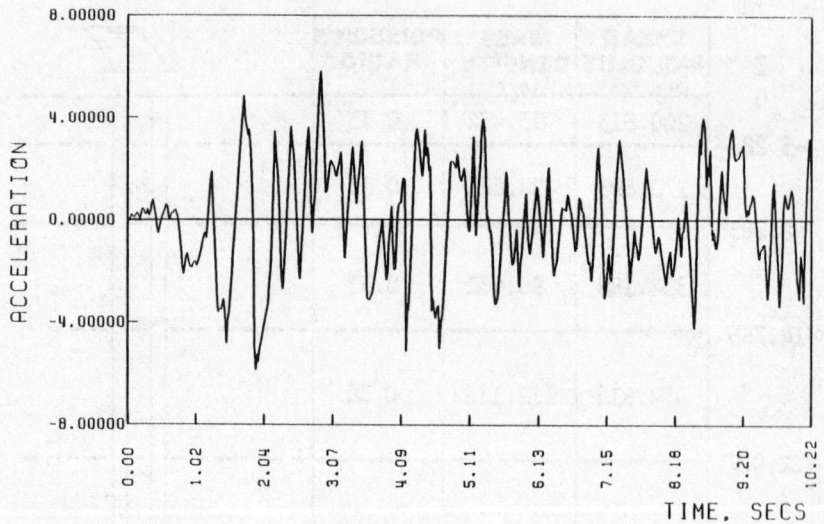

Fig. 4.25 Input earthquake motion

for the different base conditions is shown in Fig. 4.26. The flexible base case is representative of a cooling tower structure supported on a ring footing assuming that the soil–pile impedance functions approach zero, while the fixed base case means that displacements u, v, w, β_ϕ and β_θ are constrained at the base node. The foundation flexibilities of 334 piles and 167 piles lie between those of the fixed base and the flexible base. As indicated in this figure, the structure tends to move as a whole except for the fixed base case, i.e. rigid body motions occur if the base is not assumed to be fixed. Thus, in calculating the actual deformations, rigid body motions should be taken into consideration in the form given in Eq. 4.50. It is seen that the base flexibility can have a significant effect on the results. Generally speaking, the flexible base will reduce the natural frequency and increase the relative displacements. Note the similarity to Fig. 4.14, where the column supports provided the flexibility.

After elimination of rigid body effects, selected results for Cases 1, 2 and 3 demonstrate the influence of the pile foundations on the deformations and stresses in the tower. The normal displacement amplitudes for node 1, $|w_1(\omega)|$, and node 6, $|w_6(\omega)|$, are shown in Figs 4.27 and 4.28. Note how the amplitudes of the resonant peaks of $|w_1(\omega)|$ near ω_1^1 decrease with the foundation flexibility. This suggests that the damping of the system increases as the foundation becomes more flexible, because of the energy which is dissipated by radiation damping into the supporting

Dynamic analysis

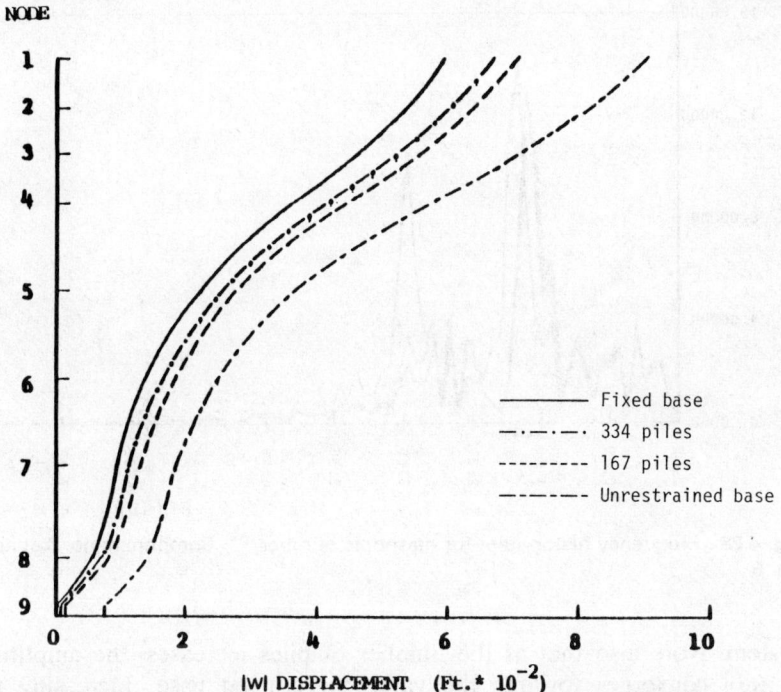

Fig. 4.26 Maximum normal displacement at $\theta = 0°$ due to earthquake

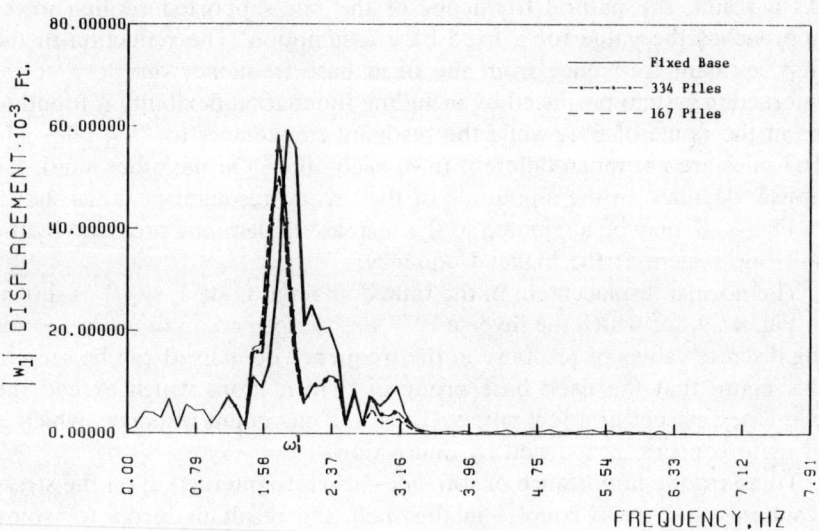

Fig. 4.27 Frequency history plot for harmonic number = 1. Component no. 3 at node no. 1

Finite element analysis of shells of revolution

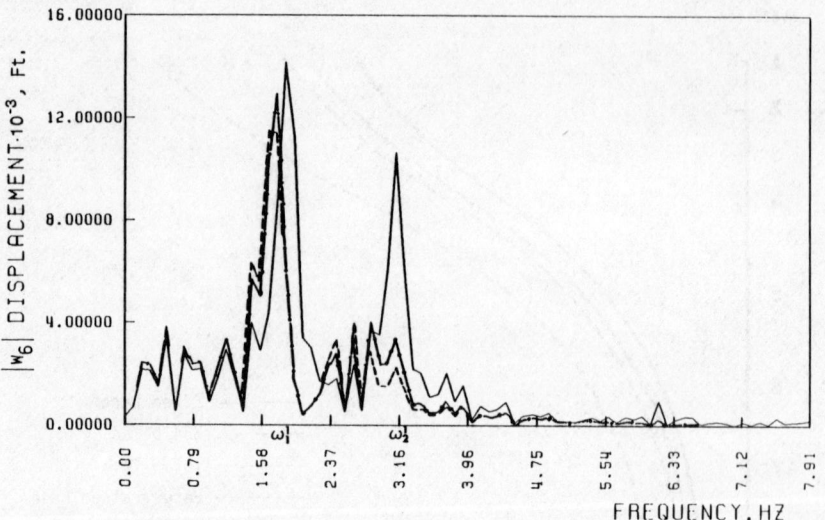

Fig. 4.28 Frequency history plot for harmonic number = 1. Component no. 3 at node no. 6

system. Note also that as the number of piles increases, the amplitude $|w_1(\omega)|$ converges towards the value for a fixed base. Increasing the number of piles effectively increases the stiffness of the foundation, primarily reducing rocking but also affecting the translation of the base. As a result, the natural frequency of the pile-supported cooling tower approaches the value for a fixed base assumption. The reduction in the first resonant frequency from the fixed base frequency due to a softer interacting system produced by including foundation flexibility is found to be in the range of 6%, while the resonant frequencies for 334 piles and 167 piles are not much different from each other. On the other hand, the drastic decrease in the amplitude of the second resonant peak, as shown in Fig. 4.28, may be attributed to the increase in damping provided by the soil–pile system at the higher frequency.

The normal displacement in the time domain at node 1, $w_1(t)$, is shown in Fig. 4.29, for which the inverse FFT was performed on the corresponding discrete values of response in the frequency domain. It can be seen in this figure that the fixed base produces deformations which exceed the pile cases, except at a few points. Thus, the maximum response, which is of main concern, is reduced by interaction.

To assess the importance of soil–pile–structure interaction on the stress resultants and stress couples in the shell, the resultant forces for some selected nodes at $\theta = 0°$ are shown in Figs 4.30 to 4.32. The sign convention follows Fig. 2.2. It may be observed that the fixed base case

Dynamic analysis

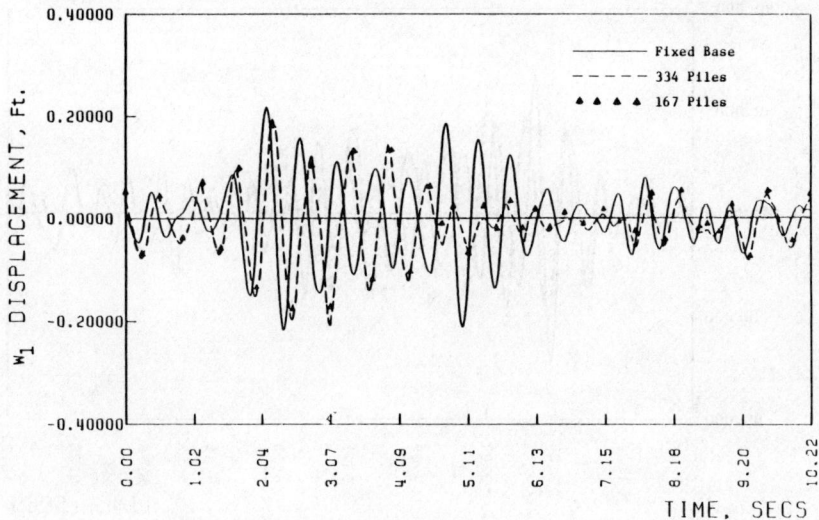

Fig. 4.29 Time history plot for harmonic number = 1. Component no. 3 at node no. 1

produces maximum resultant forces larger than for both pile cases. Since the response curves for the 167 pile case are close to those for the 334 pile case, only the peak values for the former case are recorded, with triangular symbols. The maximum responses are listed in Table 4.9. The comparison of the maximum response values shows that the interaction

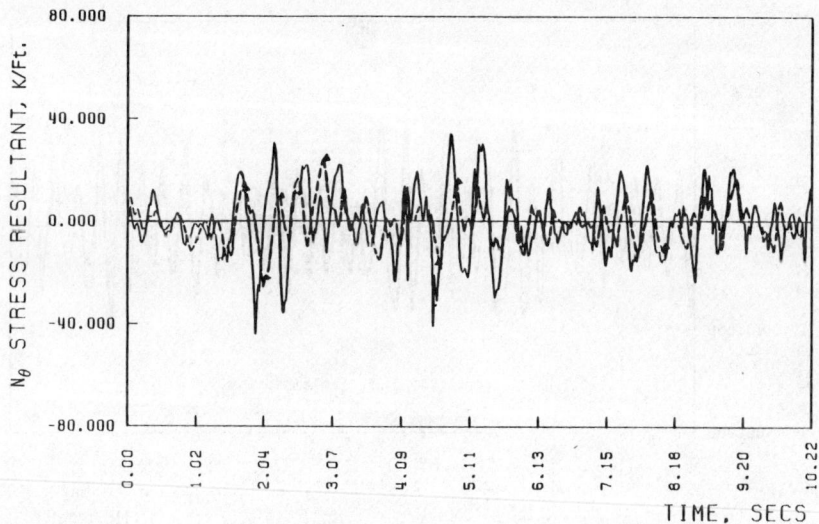

Fig. 4.30 Time history plot for harmonic number = 1. Component no. 2 at node no. 1

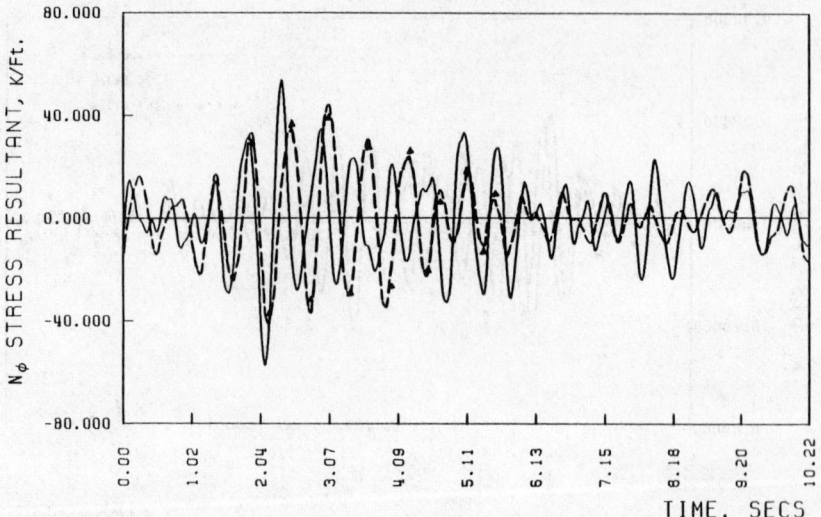

Fig. 4.31 Time history plot for harmonic number = 1. Component no. 1 at node no. 5

effect indeed reduces the stresses in the shell when a strong motion earthquake loading is applied. The percentage reduction of absolute maximum resultant forces is given in Table 4.10 in comparison with the fixed base case. Obviously, the difference between two pile cases is slight, as explained earlier. However, the reduction in N_θ at the top node may reach 44% for the 167 pile case. The reduction in N_ϕ along the shell,

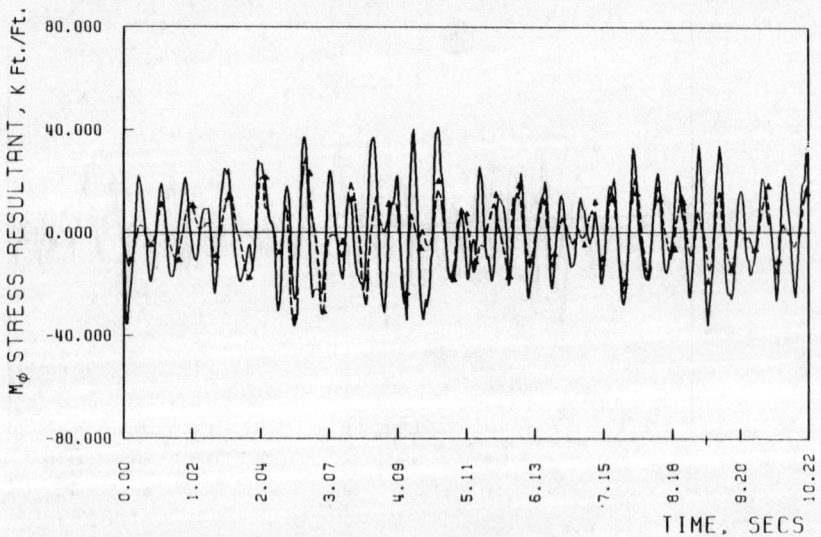

Fig. 4.32 Time history plot for harmonic number = 1. Component no. 3 at node no. 8

Table 4.9 Maximum response (at $\theta = 0°$) in the shell

Case	Maximum resultant forces and displacements*							
	N_θ Node 1	N_ϕ Node 4	N_ϕ Node 5	N_ϕ Node 7	N_ϕ Node 8	M_ϕ Node 8	M_θ Node 8	w Node 1
Fixed	+34.44 (4.82)† -43.83 (1.90)	+79.00 (5.00) -75.93 (4.76)	+54.79 (2.38) -57.14 (2.06)	+41.49 (2.38) -48.53 (2.06)	+44.62 (1.82) -53.33 (2.06)	+40.50 (4.64) -36.01 (8.70)	+8.47 (4.64) -7.04 (4.44)	+0.2176 (2.16) -0.2188 (2.36)
334 piles	+25.04 (2.96) -25.39 (2.04)	+61.06 (2.56) -64.31 (2.20)	+43.83 (3.08) -39.06 (2.20)	+33.66 (3.08) -32.88 (2.16)	+44.71 (3.08) -42.27 (8.06)	+28.96 (2.78) -35.92 (2.56)	+5.32 (2.78) -7.00 (2.56)	+0.2060 (2.16) -0.1900 (3.08)
167 piles	+24.31 (2.96) -22.35 (2.04)	+61.06 (2.56) -61.06 (2.20)	+39.21 (3.08) -36.79 (2.20)	+29.35 (3.08) -28.18 (2.16)	+41.57 (3.08) -39.80 (8.06)	+22.70 (2.78) -33.57 (2.56)	+4.38 (2.78) -6.58 (2.56)	+0.1960 (2.16) -0.1738 (3.08)

* Units: Stress resultants, k/ft; Displacement, ft; Stress couples, ft-k/ft.
† The values in parentheses are the times when the response occurs.

Finite element analysis of shells of revolution

Table 4.10 Percentage reduction of stress resultants and stress couples compared to the fixed base case

Case	Resultant forces						
	N_θ Node 1	N_ϕ Node 4	N_ϕ Node 5	N_ϕ Node 7	N_ϕ Node 8	M_ϕ Node 8	M_θ Node 8
334 piles	42	18.6	23.3	30.6	16.1	11.3	17.3
167 piles	44	22.7	31.4	39.5	22.0	17.1	22.3

which is in the range of 30% of the fixed base solution, may permit a thinner cross-section to be used and require less meridional reinforcement.

The axial force, bending moment and shear force in the columns are calculated at $\theta = 0°$ and the results are shown in Table 4.11. It may be noted that there is a sharp decrease in the shear force, while the axial force and bending moment are also decreased.

Table 4.11 Maximum column forces at $\theta = 0°$ (horizontal ground motion)

Case	Axial force (kips)	Bending moment (ft-kips)	Shear force (kips)
Fixed	862.117	654.711	746.112
334 piles	722.768	580.687	291.792
167 piles	671.975	542.650	291.467

4.6.5 Cylindrical shell under blast load

The cylindrical shell shown in Fig. 4.33 is analyzed for a $j = 0$ harmonic load, as shown in Fig. 4.34. For this harmonic, the period corresponding to the third mode is 2.420×10^{-4} s whereas, in the case of the fifth mode, it was found to be 2.415×10^{-4} s. Thus, a suitable value of the time step Δt, at one-tenth of the period, is approximately 2.5×10^{-5} s. Although the first time step cannot exceed 5×10^{-6} s because of the load peak, larger time steps may be used subsequently. The normal displacement at the top of shell, as obtained by a SHORE-III analysis and by Johnson and Grief,[23] are shown in Fig. 4.35. It may be noted that there is excellent agreement between the SHORE-III solution and Johnson and Grief's numerical integration solution.

Dynamic analysis

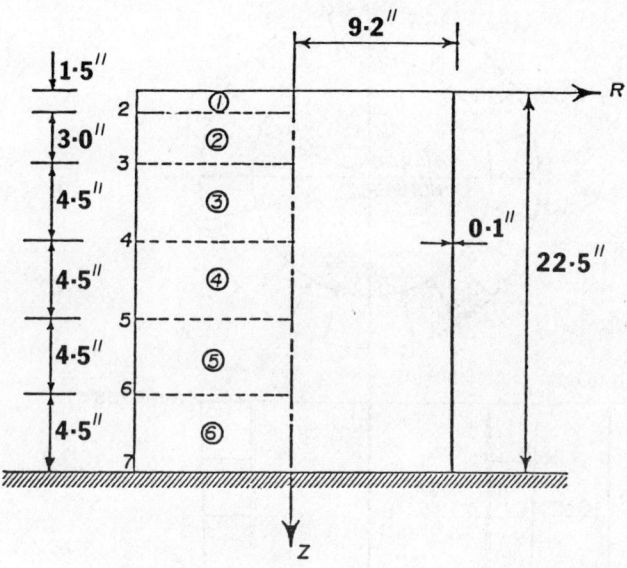

MODULUS OF ELASTICITY = 10.5×10^6 psi
POISSON'S RATIO = 0.3
MASS DENSITY = 2.4×10^{-4} lb. sec^2/in^4

Fig. 4.33 Fixed base cylindrical shell

4.6.6 Hyperboloidal shell under dynamic wind load

The shell shown in Fig. 4.10 is analyzed for the measured wind loading plotted in Fig. 4.36, based on digitized data reported at 0.5 s intervals. The pressure was assumed constant over the height of the tower, and eight harmonics, as shown in Fig. 4.36, were used. A time step of 0.025 s, about 1/50 of the fundamental period, was used with damping coefficients $\alpha_1 = 0.276$ and $\alpha_2 = 0.0058$. The resulting time history of the normal displacement, w, at the throat level and the meridional force, N_ϕ, at the base are shown in Figs 4.37 and 4.38. The SHORE-III solution was based on eleven third-order elements,[8] whereas, in Ref. 24, twenty elements were used.

A more extensive application of time history analysis is contained in Ref. 25.

4.6.7 Free vibration of fluid-filled cylindrical shell

Fluid–shell interaction is an important phenomenon in many applications. The rotational shell element derived in the preceding chapters has been

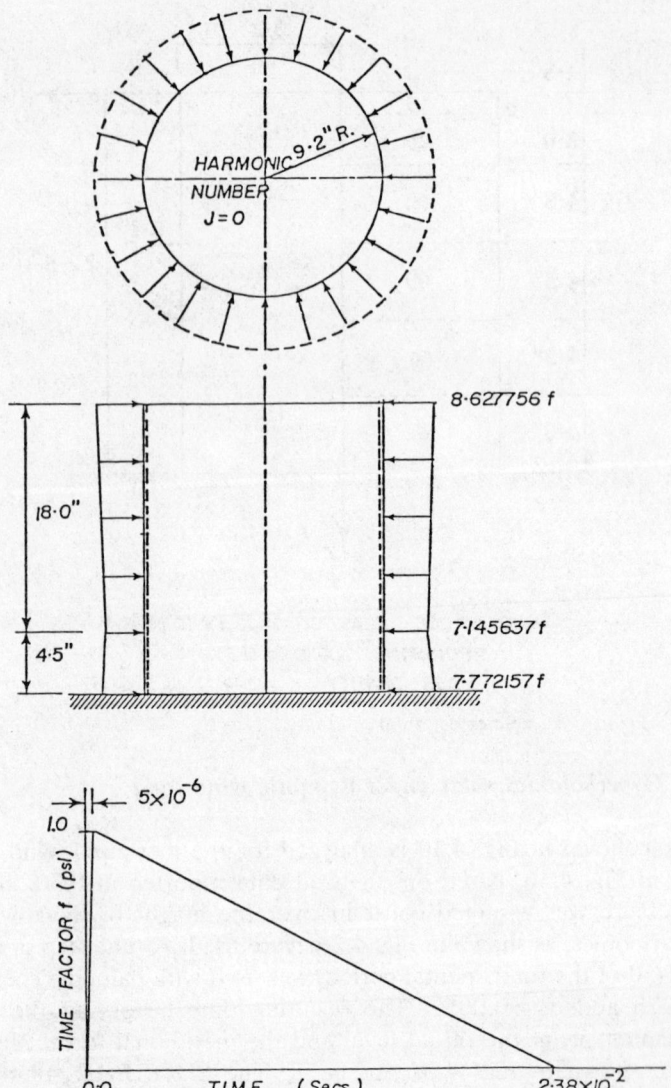

Fig. 4.34 Blast loading on a cylindrical shell

extended to include this effect.[26] The fluid is characterized as irrotational, inviscid and incompressible, and is represented by a velocity potential function which satisfies Laplace's equation. The expanded equations of motion contain the hydrodynamic pressure of the fluid with an added mass matrix.[26]

In Fig. 4.39, natural frequencies of a cylindrical shell with varying

Dynamic analysis

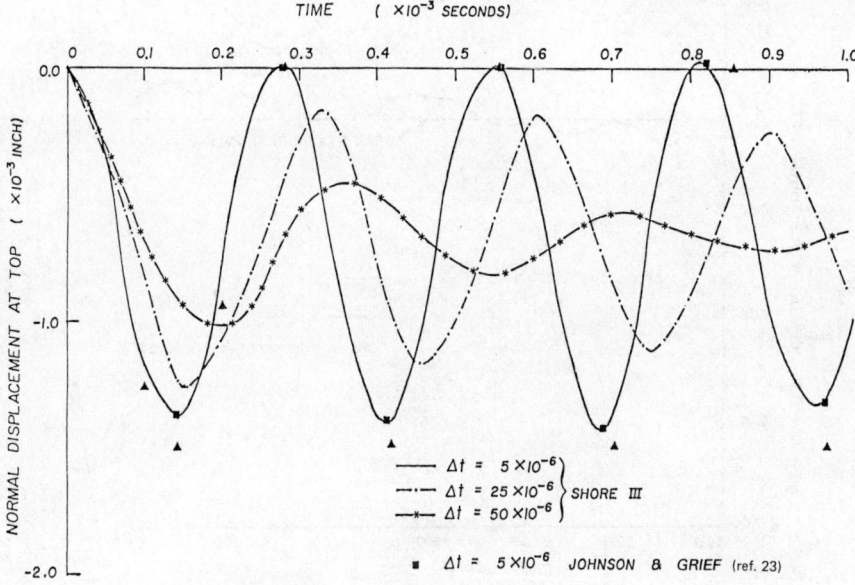

Fig. 4.35 Time history of normal displacement at the top of a cylindrical shell

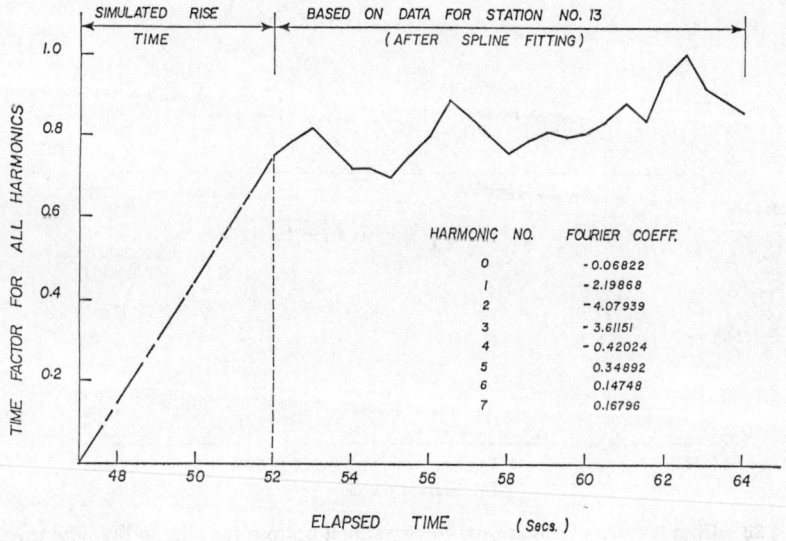

Fig. 4.36 Time history of wind loading for a hyperboloidal tower

125

Finite element analysis of shells of revolution

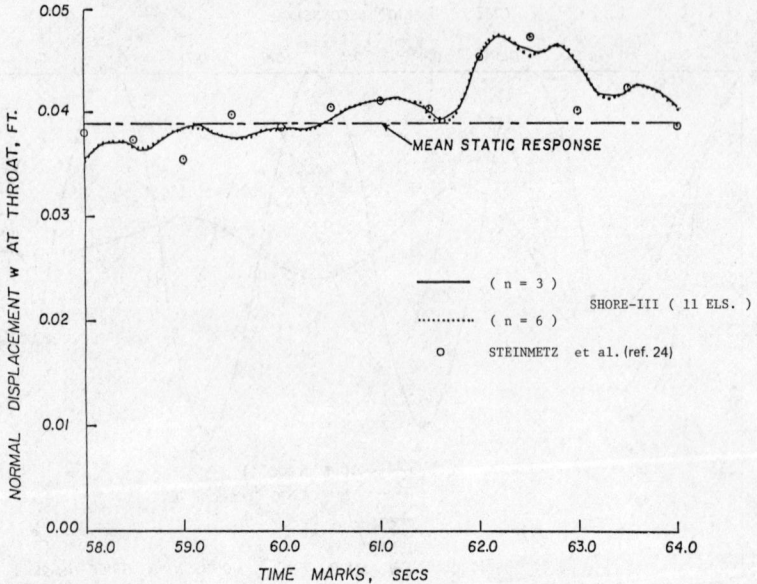

Fig. 4.37 Time history of normal displacement at throat level for a hyperboloidal tower under wind load

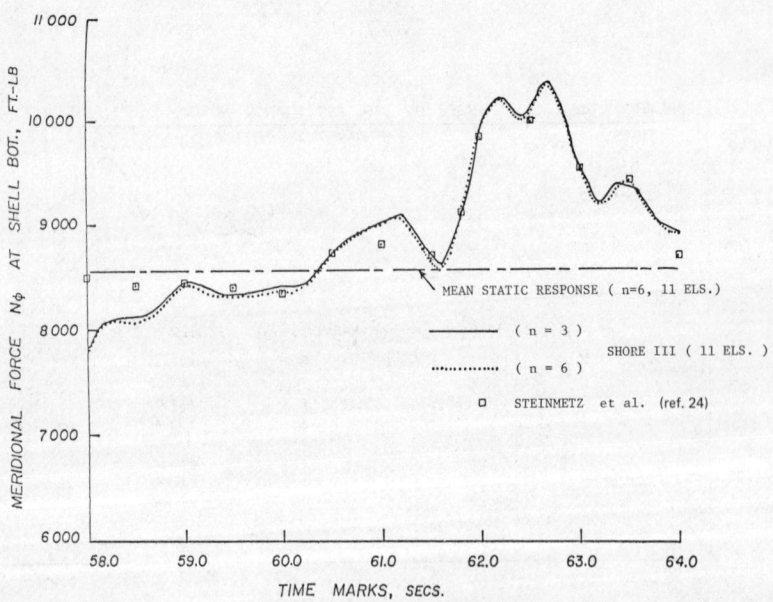

Fig. 4.38 Time history of meridional force at shell bottom for a hyperboloidal tower under wind load

Dynamic analysis

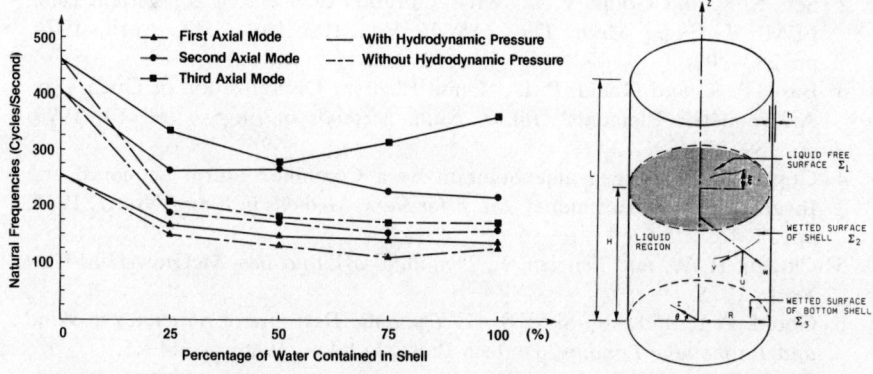

Fig. 4.39 Natural frequencies of a fluid-filled cylindrical shell. $E = 30\,000\,000$ lb in.$^{-2}$; $h = 0.5$ in.; $L = 192$ in.; $R = 60$ in.; $\rho_s = 0.000736$ lb sec^2 in.$^{-4}$; $\nu = 0.3$.

water levels are shown. Results are plotted for both the hydrodynamic pressure properly included and the fluid mass simply added to the structural mass. It is apparent that the major influence of the hydrodynamic pressure is on the higher axial modes.

Appendix: Soil properties

Case I. The soil consists of 500 ft of uniform medium sand with 75% relative density ($\gamma_d = 0.10675$ kips/cu. ft).[27] The value of the shear modulus coefficient is taken as $G = 1827$ kips/sq. ft.[28] with the value of Poisson's ratio, $\nu = 0.35$. This case is representative of a soft to intermediate soil condition.

Case II. To study the effect of a stiff and shallow soil condition, the soil depth was reduced to 250 ft. The soil is assumed to be dense sand and gravel ($\gamma_d = 0.0115$ kips/cu. ft)[27] with $G = 2675$ kips/sq. ft.[28] and $\nu = 0.4$. This case is representative of an intermediate to stiff soil condition.

Case III. This case is formulated such that fundamental frequency of the soil layer is close to that of the structure. Strong amplification due to resonance effects, if present, would show up. The soil is assumed to be stiff clay ($\gamma_d = 0.13$ kips/cu. ft)[27] with a depth of 600 ft. The values of the shear modulus coefficient and Poisson's ratio are taken as $G = 2315$ kips/sq. ft[28] and $\nu = 0.4$, respectively.

Case IV. The structure is directly founded on competent rock and, therefore, the soil–structure interaction effect is negligible. This case represents an important convergence point for the solution technique.

References

1. Geradin, M., 'Error Bounds for Eigenvalues Analysis by Elimination of Variables', *J. of Sound and Vibration*, Vol. 19, 1971, pp. 111–132.

2. Sen, S. K. and Gould, P. L., 'Free Vibration of Shells of Revolution using FEM', *J. Engrg Mech. Div., ASCE*, Vol. 100, No. EM2, April, 1974, pp. 283–303.
3. Basu, P. K. and Gould, P. L., 'Finite Element Discretization of Open-Type Axisymmetric Elements', *Int. J. Num. Methods in Engrg*, Vol. 14, 1979, pp. 159–178.
4. Gupta, K. K., 'Eigenvalue Solution by a Combined Sturm Sequence and Inverse Iteration Technique', *Int. J. for Num. Methods in Engrg*, Vol. 7, 1973, pp. 17–42.
5. Clough, R. W. and Penzien, J., *Dynamics of Structures*, McGraw-Hill, New York, 1975.
6. Gould, P. L. and Abu-Sitta, S. H., *Dynamic Response of Structures to Wind and Earthquake Loading*, Pentech Press, London, 1980, pp. 44–45.
7. Gould, P. L., Sen, S. K. and Suryoutomo, H., 'Dynamic Analysis of Column-Supported Hyperboloidal Shells', *Int. J. Earthquake Engineering and Structural Dynamics*, Vol. 2, 1974, pp. 269–279.
8. Basu, P. K. and Gould, P. L., 'SHORE-III Theoretical Manual', Research Report No. 48, Structural Division, Dept. of Civil Engineering, Washington University, St. Louis, Mo, Sept. 1977, pp. 20–22.
9. Weingarten, V. I., 'Free Vibration of Thin Cylindrical Shells', *AIAA J.*, Vol. 2, No. 4, April 1964, pp. 717–722.
10. Carter, R. L., Robinson, A. R. and Schnobrich, W. C., 'Free and Forced Vibrations of Hyperboloidal Shells of Revolution', Structural Research Series No. 334, University of Illinois, Urbana, February, 1968.
11. Kraus, H., *Thin Elastic Shells*, John Wiley and Sons, 1967.
12. Kalnins, A. and Kraus, H., 'Effect of Transverse Shear and Rotary Inertia on Vibration of Spherical Shells', *Proc. 5th U.S. National Congress of Applied Mechanics, ASME, 1966*, p. 134.
13. Gould, P. L., Suryoutomo, H. and Sen, S. K., 'Stresses in Column-Supported Hyperboloidal Shells Subject to Seismic Loading', *Int. J. Earthquake Engineering and Structural Dynamics*, Vol. 5, 1977, pp. 3–14.
14. Fagel, L. W. and Liu, S. C., 'Earthquake Interaction for Multi-Story Buildings', *J. Engrg Mech. Div., ASCE*, Vol. 98, No. EM4, Aug. 1972, pp. 929–945.
15. El-Shafee, O. M. and Gould, P. L., 'Cooling Tower Founded on Soil', *J. Struct., Div. ASCE*, Vol. 108, No. ST4, April, 1982, pp. 800–813.
16. El-Shafee, O. M. and Gould, P. L., 'Dynamic Axisymmetric Soil Model for a Flexible Ring Footing', *Int. J. Earthquake Engineering and Structural Dynamics*, Vol. 8, 1980, pp. 479–498.
17. El-Shafee, O. M. and Gould, P. L., 'Analytical Method for Determining Seismic Response of Cooling Towers on Footing Foundation', Research Report No. 55, Structural Division, Washington University, St. Louis, Mo., Sept. 1979.
18. Pandya, V. and Setlur, A. Y., 'Seismic Soil–Structure Interaction by Finite Elements: Case Studies', *Proceedings of the 2nd ASCE Speciality Conference on Structural Design of Nuclear Plant Facilities*, Vol. 1-A, Dec. 1975, pp. 826–848.

19. Gould, P. L., *Static Analysis of Shells*, Lexington Books, D. C. Heath and Co., New York, NY, 1977.
20. Shye, K. Y. and Robinson, A. R., 'Dynamic Soil-Structure Interaction', Dept. of Civil Engineering, University of Illinois at Urbana-Champaign, Structural Research Series No. 464, UILU-ENG-80-2018, Sept., 1980.
21. Lee, B. J. and Gould, P. L., 'Complex Response Analysis of Shells of Revolution Including Uniform Base Translation and Rocking', *J. Earthquake Engineering and Structural Dynamics*, in press.
22. Lee, B. J. and Gould, P. L., 'Seismic Response of Pile Supported Cooling Towers', *J. Struct. Engrg, ASCE*, Vol. 12, 1984, pp. 507–519.
23. Johnson, D. E. and Grief, R., 'Dynamic Response of a Cylindrical Shell: Two Numerical Methods', *AIAA J.*, Vol. 4, No. 3, March, 1966, pp. 486–494.
24. Steinmetz, R. L., Billington, D. D. and Abel, J. F., 'The Dynamic Response of Hyperbolic Cooling Towers to Wind', Research Report No. 76-SM-5, Princeton University, April 1976.
25. Basu, P. K. and Gould, P. L., 'Cooling Towers Using Measured Wind Data', *J. Struct. Div. ASCE*, Vol. 106, No. 573, March 1980, pp. 579–600.
26. Phan, L. T. and Gould, P. L., 'Dynamic Effects of Liquid in Liquid-Shell Systems', Engineering Mechanics in Civil Engineering, *Proc. Fifth ASCE EMD Specialty Conference*, Laramie, Wyoming, Aug. 1984, pp. 281–284.
27. Lambe, T. W. and Whitman, R. V., *Soil Mechanics*, John Wiley and Sons, New York, NY, 1969, p. 31.
28. Seed, H. B. and Idriss, I. M., 'Soil Moduli and Damping Factors for Dynamic Response Analysis', Report No. EERC 70-10, University of California, Berkeley, Ca, Dec. 1970.

5 Geometric nonlinearity and instability

5.1 General

A generalization of the thin shell formulation to incorporate geometric nonlinearities is required to perform instability and finite deformation analyses. The focus for this generalization is the strain–displacement relations, which ultimately leads to a modified set of governing equations. Subsequently, the equations may be simplified through the elimination of terms that are presumed to be insignificant. This is often carried out with regard to the objectives of addressing a specific class of problems, so that there are several varieties of nonlinear equations for shells in the literature.

In this brief development, the Fourier series' expressions in the circumferential variable are retained and transverse shearing strains are included. Thus, the formulation is consistent with the linear theory, in so far as possible. Since harmonic coupling is not addressed, however, the generality of the formulation is restricted.

5.2 Strain–displacement relations

The linear strain–displacement relations in harmonic form were given as Eq. 2.30 with the strains defined in Eqs 2.20–2.22.

The nonlinear geometric terms which arise are of three types:[1] (1) Products of two displacements (or derivatives thereof); (2) Products of one displacement and one rotation; and (3) Products of two rotations. The most frequently considered case is the latter, for which the corresponding relations were given as Eqs 2.34–2.44.

As mentioned in Section 5.1, there are several varieties of nonlinear strain–displacement equations. If the transverse shearing strains γ^j_ϕ and γ^j_θ are suppressed, the expressions for the in-plane strains $\hat{\varepsilon}^j_\phi$, $\hat{\varepsilon}^j_\theta$ and $\hat{\varepsilon}^j_{\theta\phi}$ reduce to the form given by Stricklin.[2] However, in the latter reference, the nonlinear contributions to the change in curvature, $\hat{\kappa}^j_\phi$, $\hat{\kappa}^j_\theta$ and $\hat{\kappa}^j_{\theta\phi}$, are not included. On the other hand, Klein[3] presented a consistent derivation which included these terms, but not the transverse shearing strains. More general relations than those utilized here are given by Novozhilov.[1]

Geometric nonlinearity and instability

The case for including transverse shearing strains, and thus providing C^0 continuity, is primarily based on the relatively small increase in computational effort required to eliminate derivatives as nodal variables. Then, there is the traditional notion that unnecessary simplifications should be avoided, even if no significant change in results is anticipated for most problems. The inclusion of the nonlinear changes in curvature in the modified kinematic equations is similarly motivated. Instability analysis, in which the prebuckled state is closely represented by the membrane theory solution, apparently may be performed adequately using only the extensional nonlinear terms.[2] More general equations than those presented here might be required in the case of finite deformation analyses of flexible shells. Of course, as deformations become significant, material nonlinearity may become a factor, as well.

The nonlinear rotations $\{\check{D}_i^j(s)\}$, used in Eqs 2.34–2.44, are related to the nodal displacements $\{\Delta_i^j\}$, Eq. 3.14, by extending Eq. 2.42 to

$$\{\check{D}_i\}^j = \lceil \Theta_4^i \rfloor \{\check{D}_i^j(s)\} = [Z_i(s)] \lceil \Theta_6^i \rfloor \{\Delta_i^j\} \tag{5.1}$$

in which

$$\lceil \Theta_6^i \rfloor = \lceil \Theta_1^i \Theta_1^j \rfloor \tag{5.2}$$

where $\lceil \Theta_1^i \rfloor$ is defined in Eq. 2.18 and $[Z_i(s)]$ is a 4×10 interpolation matrix of the type used in Section 3.1.2. The precise form of $[Z_i(s)]$ for this equation will be discussed subsequently.

5.3 Modified equilibrium equations

The generalization of the static equilibrium equations is accomplished by re-evaluating the total potential energy, Eqs 2.51 and 2.56, with the nonlinear strains $\{\check{\varepsilon}\}$, as defined by Eq. 2.34, added to $\{\varepsilon\}$, which was used in Section 2.7.1. After writing the Fourier components

$$\{\hat{\varepsilon}\}^j = \{\varepsilon\}^j + \{\check{\varepsilon}\}^j \tag{5.3}$$

in terms of the displacements via Eqs 2.20–2.22 and 2.30 for $\{\varepsilon\}^j$ and Eqs 2.35–2.41 for $\{\check{\varepsilon}\}^j$, the resulting expression may be separated into three distinct groups:[4]

(1) Terms which are products of the components of $\{\varepsilon\}^j$.
(2) Terms which are the product of a term from $\{\varepsilon\}^j$ and a term from $\{\check{\varepsilon}\}^j$.
(3) Terms which are products of the components of $\{\check{\varepsilon}\}^j$.

The terms in group (1) correspond to an initial state of equilibrium which is presumed to be stable, thus satisfying Eq. 2.59 independently.

Then, groups (2) and (3) may be interpreted as a result of a perturbation of the equilibrium state (1).

If the terms of (1) are regarded as constant during the perturbation, then the group (2) terms form a second equilibrium state under the initial stress state corresponding to (1).[4]

Finally, the group (3) terms are purely the strains due to the perturbation. Since, the entire system is in equilibrium and the terms of groups (1) and (2) independently lead to states of equilibrium, the terms of group (3) provide another independent equilibrium condition which may be interpreted as the vanishing of the second variation of the total potential energy functional.[4] This is expressed through the condensed element equation

$$([\bar{k}_i^j] + \bar{\lambda}^i [\bar{k}_{\sigma_i}^j]) d\{\Delta_i^j\} = d\{\bar{\bar{\mathscr{F}}}_i^j\} \tag{5.4}$$

in which $d\{\Delta_i^j\}$ = the incremental displacement vector; $d\{\bar{\bar{\mathscr{F}}}_i^j\}$ = the incremental load vector; $[\bar{k}_{\sigma_i}^j]$ = the element geometric stiffness matrix, which comes from the group (3) terms; and $\bar{\lambda}^i$ is a scalar which is proportional to the magnitude of the applied loading and therefore scales the state of stress prior to the perturbation. Although Eq. 5.4 is the classical form, a slight generalization to include follower terms will be discussed in Section 5.7.

For our purposes it is necessary to define $[\bar{k}_{\sigma_i}^j]$ more specifically.

5.4 Geometric stiffness matrix

5.4.1 Definition

The geometric stiffness matrix $[\bar{k}_{\sigma_i}^j]$ is defined with respect to $d\{\Delta_i^j\}$ by applying the general approach of Zienkiewicz[5] to the shell of revolution geometry and neglecting harmonic coupling.[6] For element i and harmonic j,

$$[\bar{k}_{\sigma_i}^j] d\{\Delta_i^j\} = \int_{\mathscr{S}} d[B_i^j]^T \{\check{N}_i^j\} \, d\mathscr{S} \tag{5.5}$$

in which $[B_i^j]$ is the transformation between the differential nonlinear strains and nodal displacements for element i, which will be developed subsequently, and $\{\check{N}_i^j\}$ is the vector of specified initial stresses. In this formulation, $\{\check{N}_i^j\}$ is assumed to include only symmetric (proportional to $\cos j\theta$) terms, which would seem to limit the application to the $j = 0$ loading condition. However, in conjunction with perhaps the most prominent application of this formulation, linear buckling analysis, Cole, Abel and Billington[7] have suggested that the initial stress prior to buckling under a non-symmetrical loading condition can be taken as the stress state along a 'critical' meridian, which is then assumed to be axisymmetric.[7] This approximation extends the applicability of the theory considerably and is said to be conservative.

Also, with respect to Eq. 5.5, the element of surface area $d\mathscr{S}$ was defined in Eq. 3.24.

5.4.2 Transformation matrix

To derive $[B_i^j]$, we consider Eq. 2.35 for harmonic j

$$\{\check{\varepsilon}_i\}^j = \tfrac{1}{2}[\check{A}_i]^j\{\check{D}_i\}^j \tag{5.6}$$

and take the differential

$$d\{\check{\varepsilon}_i\}^j = \tfrac{1}{2}[[\check{A}_i]^j d\{\check{D}_i\}^j + d[\check{A}_i]^j\{\check{D}_i\}^j] \tag{5.7}$$

The second term of Eq. 5.7 may be written as

$$d[\check{A}_i]^j\{\check{D}_i\}^j = [\hat{A}_i]^j d\{\check{D}_i\}^j \tag{5.8}$$

in which

$$[\hat{A}_i]^j = [\Theta_3^j][\hat{A}_i^j(s)] \tag{5.9}$$

$$[\hat{A}_i^j(s)]^\mathrm{T} = \begin{bmatrix} \check{\beta}_\phi^j & 0 & \check{\beta}_\theta^j & \dfrac{2}{R_\phi}\beta_\phi^j & 0 & \dfrac{2}{R_\theta}\beta_\theta^j & 0 & 0 \\ 0 & \check{\beta}_\theta^j & \check{\beta}_\phi^j & 0 & \dfrac{2}{R_\theta}\beta_\theta^j & \dfrac{2}{R_\phi}\beta_\phi^j & 0 & 0 \\ 0 & 0 & 0 & 0 & 0 & 0 & 0 & 0 \\ 0 & 0 & 0 & 0 & 0 & 0 & 0 & 0 \end{bmatrix}_{4\times 8} \tag{5.10}$$

and $[\Theta_3^j]$ is defined in Eq. 2.37. Then, Eq. 5.7 may be rewritten as

$$d\{\check{\varepsilon}_i\}^j = [\hat{A}_i]^j d\{\check{D}_i\}^j \tag{5.11}$$

in which

$$[\hat{A}_i]^j = \tfrac{1}{2}[[\check{A}_i]^j + [\hat{A}_i]^j] \tag{5.12}$$

Now, we substitute the differential of Eq. 5.1 into Eq. 5.11 to get

$$d\{\check{\varepsilon}_i\}^j = [\hat{A}_i]^j[Z_i(s)][\Theta_6^j] d\{\Delta_i^j\} \tag{5.13}$$

Finally, the transformation between the differential nonlinear strains and nodal displacements is established as

$$[B_i^j] = [\hat{A}_i]^j[Z_i][\Theta_6^j] \tag{5.14}$$

5.4.3 Evaluation

From the form of Eq. 5.14, it is obvious that the Fourier coefficients of the displacement vector serve as the nodal variables.

Next, we consider

$$d[B_i^j]^\mathrm{T} = [\Theta_6^j][Z_i]^\mathrm{T} d[\hat{A}_i]^{j\mathrm{T}} \tag{5.15}$$

whereupon Eq. 5.5 becomes

$$[\bar{k}_{\sigma_i}^j]d\{\Delta_i^j\} = \int_{\mathscr{S}} [\Theta_6^j][Z_i]^\mathrm{T} d[\hat{A}_i]^{j\mathrm{T}}\{\check{N}_i^j\}\,d\mathscr{S} \tag{5.16}$$

Finite element analysis of shells of revolution

The last two terms on the r.h.s. of Eq. 5.16 may be rearranged as[6]

$$d[\hat{A}]^{j^T}\{\check{N}_i^j\} = [\check{N}_i^j]d\{\check{D}_i\}^j \qquad (5.17)$$

in which

$$[\check{N}_i^j] = \begin{bmatrix} N_\phi^j & & & \\ 0 & N_\theta^j & \text{(Symmetric)} & \\ \dfrac{M_\phi^j}{R_\phi} & 0 & 0 & \\ 0 & \dfrac{M_\theta^j}{R_\theta} & 0 & 0 \end{bmatrix} \qquad (5.18)$$

is the 'axisymmetric' prebuckled stress matrix. Upon substitution of Eq. 5.1 into 5.17, we obtain

$$d[\hat{A}_i]^{j^T}\{\check{N}_i^j\} = [\check{N}_i^j][Z_i]\lceil \Theta_6^j \rfloor d\{\Delta_i^j\} \qquad (5.19)$$

Introducing Eq. 5.19 into Eq. 5.16, noting Eq. 3.14, and equating factors of $d\{\Delta_i^j\}$, we have

$$[\bar{k}_{\sigma_i}^j] = L_i \int_0^1 \int_{-\pi}^{\pi} \lceil \Theta_6^j \rfloor [Z_i]^T [\check{N}_i^j][Z_i] \lceil \Theta_6^j \rfloor R \, d\theta \, ds \qquad (5.20)$$

Since $[\check{N}_i^j]$ is based on a symmetric stress state, only $\lceil \Theta_6^j \rfloor$ is a function of θ and the integral of Eq. 5.20 contains either $\sin^2 j\theta$, $\cos^2 j\theta$ or $\sin j\theta \cos j\theta$ in each term. Therefore,

$$\int_{-\pi}^{\pi} \lceil \Theta_6^j \rfloor [Z_i]^T [\check{N}_i^j][Z_i] \lceil \Theta_6^j \rfloor d\theta = \lambda \pi [Z_i]^T [\check{N}_i^j][Z_i] \qquad (5.21)$$

in which

$$\lambda = 2 \text{ for } j = 0 \quad \text{and} \quad \lambda = 1 \text{ for } j = 1$$

Finally the geometric stiffness matrix is written as

$$[\bar{k}_{\sigma_i}^j] = \lambda \pi L_i \int_0^1 [Z_i]^T [\check{N}_i^j][Z_i] R \, ds$$

$$= \lambda \pi L_i \int_0^1 [\check{k}_{\sigma_i}^j] R \, ds \qquad (5.22)$$

The actual integration is best accomplished by a numerical scheme such as Gaussian quadrature.

Since $[\bar{k}_{\sigma_i}^j]$ is obviously proportional to $[\check{N}_i^j]$, it is sometimes called the initial stress stiffness matrix. It should be noted that if the nonlinear terms in the change in curvature–displacement equations are dropped, as discussed in Sections 2.4.3 and 5.2, the prebuckled stress matrix will only contain N_ϕ^j and N_θ^j and no stress couple terms.

Geometric nonlinearity and instability

5.5 Displacement interpolation functions

5.5.1 General considerations

In order to evaluate $[\bar{k}_{\sigma_i}^j]$, appropriate displacement interpolation functions $[Z_i(s)]$ need to be selected. A general form for such functions was introduced in Section 3.1.2 and refined in Section 3.2.1. However, since we are now concerned with nonlinear displacements, it is of interest to re-examine this topic briefly.

In the solution of geometrically nonlinear shells of revolution under both static and dynamic loading, Stricklin[2] used different order interpolation functions for the linear and nonlinear parts. In that formulation, transverse shearing strains were suppressed, which resulted in *second* derivatives of the displacements entering the linear kinematic law (as opposed to only first derivatives in Eq. 2.31). In turn, a minimum of C^1 continuity was required, which is achievable only by quadratic or higher order interpolation functions. The corresponding nonlinear relations given in Ref. 2 contain only *first* derivatives of the nodal displacement (as opposed to *no* derivatives in Eq. 5.11). Thus, only minimum C^0 continuity was required, which can be satisfied by linear interpolation functions. Based on these requirements and reflecting a notion prevalent in early finite element methodology, namely that low-order interpolations are preferable because of simplicity, second-order displacement functions were chosen for the linear displacements while first order polynomials were used for the nonlinear displacements in Ref. 2.

If the somewhat less prevalent, but still respectable, notion of employing the lowest order displacements consistent with minimum continuity requirements is accepted, the present formulation, in which transverse shearing strains are included, is attractive in that first-order expansions can seemingly be used for both the linear and nonlinear displacements. However, in view of the numerical results presented in the preceding chapters, it is doubtful that this would be the most efficient procedure. Rather, the employment of high-order displacement functions is suggested. Nevertheless, in the absence of contradictory numerical evidence for nonlinear problems, it is possible that first order interpolations for the linear displacements may be desirable in some applications and appropriate interpolation functions are developed here for this, as well as the general, case.

5.5.2 First-order interpolations

The first order interpolation function is generated from sub-matrix $[\bar{Z}_{ia}]$ of matrix $[\bar{Z}_i]$, as defined in Eqs 3.19 and 3.20.[6]

Finite element analysis of shells of revolution

The transverse shearing strains γ_ϕ^j and γ_θ^j are expressed in terms of the nodal variables defined in Eqs 2.19 by the last two expressions in Eq. 2.31:

$$\gamma_{\phi i}^j = -\frac{1}{R_\phi} u_j^i + \frac{1}{L_i} w_{i,s}^j + \beta_{\phi i}^j \qquad (5.23)$$

and

$$\gamma_{\theta i}^j = -\frac{\sin\phi}{R} v_i^j - \frac{j}{R} w_i^j + \beta_{\theta i}^j \qquad (5.24)$$

Next, the interpolations for the nonlinear displacements $\{\check{D}_i^j(s)\}$, as defined in $\{\check{D}^j(s)\}$ (Eq. 2.44), are developed from $[\bar{Z}_{ia}]$. For $\beta_{\phi i}^j$ and $\beta_{\theta i}^j$, the fourth and fifth rows are used directly. For $\bar{\beta}_{\phi i}^j$, following Eq. 5.23, $1/R_\phi$ times the first row minus the third row is used; for $\bar{\beta}_{\theta i}^j$, following Eq. 5.24, $\sin\phi/R$ times the second row plus j/R times the third row is used. The result is[6]

$$[Z_i]^T = [Z_{ia}]^T = \begin{bmatrix} \dfrac{1-s}{R_\phi} & 0 & 0 & 0 \\ 0 & \dfrac{\sin\phi}{R}(1-s) & 0 & 0 \\ -\dfrac{1}{L_i} & \dfrac{j}{R}(1-s) & 0 & 0 \\ 0 & 0 & 1-s & 0 \\ 0 & 0 & 0 & 1-s \\ \dfrac{s}{R_\phi} & 0 & 0 & 0 \\ 0 & \dfrac{\sin\phi}{R}s & 0 & 0 \\ \dfrac{1}{L_i} & \dfrac{j}{R}s & 0 & 0 \\ 0 & 0 & s & 0 \\ 0 & 0 & 0 & s \end{bmatrix} \qquad (5.25)$$

which may be used directly in Eq. 5.22 and the preceding equations.

As we noted earlier, an attraction of low-order polynomials is simplicity. The integrand for the geometric stiffness matrix, Eq. 5.22, is easily written from Eqs 5.18 and 5.25:

$$[\check{k}_{\sigma_i}^j] = \begin{bmatrix} I & II^T \\ II & III \end{bmatrix} \qquad (5.26)$$

in which

$$I = \begin{bmatrix} \left(\dfrac{1-s}{R_\phi}\right)^2 N_\phi^i & & & & & \\ 0 & \left(\dfrac{1-s}{R_\theta}\right)^2 N_\theta^j & & \text{(Symmetric)} & & \\ -\dfrac{(1-s)}{R_\phi L_i} N_\phi^j & \dfrac{(1-s)^2}{R_\theta R} jN_\theta^j & \left[\left(\dfrac{1}{L_i}\right) N_\phi^j + \left(\dfrac{1-s}{R}\right)^2 j^2 N_\theta^j\right] & & & \\ \left(\dfrac{1-s}{R_\phi}\right)^2 M_\phi^j & 0 & -\dfrac{(1-s)}{R_\phi L_i} M_\phi^j & 0 & & \\ 0 & \left(\dfrac{1-s}{R_\theta}\right)^2 M_\theta^j & \dfrac{(1-s)^2}{R_\theta R} jM_\theta^j & 0 & 0 & \end{bmatrix}$$

$$II = \begin{bmatrix} \dfrac{s(1-s)}{R_\phi^2} N_\phi^j & 0 & -\dfrac{s}{R_\phi L_i} N_\phi^j & \dfrac{s(1-s)}{R_\phi^2} M_\phi^j & 0 \\ 0 & \dfrac{s(1-s)}{R_\theta^2} N_\theta^j & \dfrac{s(1-s)}{R_\theta R} jN_\theta^j & 0 & \dfrac{s(1-s)}{R_\theta^2} M_\theta^j \\ \dfrac{(1-s)}{R_\phi L_i} N_\phi^j & \dfrac{s(1-s)}{R_\theta R} jN_\theta^j & \left[-\left(\dfrac{1}{L_i}\right)^2 N_\phi^j + \dfrac{s(1-s)}{R^2} j^2 N_\theta^j\right] & \dfrac{(1-s)}{R_\phi L_i} M_\phi^j & \dfrac{s(1-s)}{R_\theta R} jM_\theta^j \\ \dfrac{s(1-s)}{R_\phi^2} M_\phi^j & 0 & -\dfrac{s}{R_\phi L_i} M_\phi^j & 0 & 0 \\ 0 & \dfrac{s(1-s)}{R_\theta^2} M_\theta^j & \dfrac{s(1-s)}{R_\theta R} jM_\theta^j & 0 & 0 \end{bmatrix}$$

$$III = \begin{bmatrix} \left(\dfrac{s}{R_\phi}\right)^2 N_\phi^j & & & & & \\ \dfrac{s}{R_\phi L_i} N_\phi^j & \left(\dfrac{s}{R_\theta}\right)^2 N_\theta^j & & \text{(Symmetric)} & & \\ 0 & \dfrac{s^2}{R_\theta R} jN_\theta^j & \left[\left(\dfrac{1}{L_i}\right)^2 N_\phi^j + \left(\dfrac{s}{R}\right)^2 j^2 N_\theta^j\right] & & & \\ \left(\dfrac{s}{R_\phi}\right)^2 M_\phi^j & 0 & \dfrac{s}{R_\phi L_i} M_\phi^j & 0 & & \\ 0 & \left(\dfrac{s}{R_\theta}\right)^2 M_\theta^j & \dfrac{s^2}{R_\theta R} jM_\theta^j & 0 & 0 \end{bmatrix}$$

5.5.3 Higher-order interpolations

Using the entire matrix $[\bar{Z}_i]$ from Eqs 3.19–3.21 and following identical steps to those outlined in the preceding section, we may develop

$$[Z_i] = [Z_{ia} \mid Z_{ib}] \tag{5.27}$$

from which the geometric stiffness matrix, Eq. 5.22, can be derived.[6] The expressions are lengthy and are omitted here.

5.6 Bifurcation buckling

5.6.1 Procedure

Numerical applications of geometric nonlinear analysis are not presented here. However, it may be instructive to indicate how the preceding equations can be utilized in perhaps the most prominent application—bifurcation buckling analysis. The global equations are assembled from the homogeneous part of Eq. 5.4 and the appropriate boundary conditions are then applied. This leads to the classical form of the buckling equations

$$([\bar{K}^j] + \bar{\lambda}^j [\bar{K}_\sigma^j]) d\{\Delta^j\} = \{0\} \tag{5.28}$$

in which $[\bar{K}_\sigma^j]$ is the global geometric stiffness matrix. As mentioned previously, some limitations on Eq. 5.28 will be discussed in Section 5.7.

The scaling factor $\bar{\lambda}^j$, defined in Section 5.3, becomes the eigenvalue and, corresponding to each $\bar{\lambda}^j$, is an eigenvector $d\{\Delta^j\}$. These are equivalent to the natural frequencies and mode shapes arising from the free vibration problem treated in Chapter 4 and, thus, the same computer code can be applied for instability analysis with the mass matrix replaced by the geometric stiffness matrix. A very detailed discussion of the computational aspects of instability analysis is presented in Ref. 8.

A word about the method of computing the prebuckling stresses in $[\check{N}_i^j]$, Eq. 5.18, which ultimately affects $[\bar{K}_\sigma^j]$, may be helpful. The form of the latter equation suggests a full bending theory analysis, since stress couples are contained. However, in the literature, it is common to find reference to the membrane theory stresses as the prebuckled stress state. If the finite element approach described in this book is to be used for the instability analysis, then it requires little more effort to perform a linear bending analysis for the prebuckled stress state. The reader is advised to take note of the method of evaluating the prebuckled stress state when interpreting results from computer analyses and from the literature.[8]

The lowest actual buckling level is determined by solving the eigenvalue equation for each harmonic until the minimum is found. Analogous to the free vibration problem, Fig. 4.5, the lowest eigenvalue generally does *not* correspond to $j = 0$, but to a much higher harmonic.

A fuller development of the modified equilibrium equations and some applications to shell buckling are contained in Ref. 4. One interesting comparison presented there contrasts various computerized methods for determining the buckling load of axially loaded cylinders with thickness h, as a function of the length L to radius R ratio. This provides a somewhat independent verification of the finite element approach. For a problem studied which has $R/h = 100$, the total length of the shell is insignificant beyond $L/R = 0.5$, indicating buckling into several axial waves. Also the method of prebuckling stress computation (membrane or bending) is shown to have a noticeable, but numerically small, effect on the buckling load for the axially loaded cylinders.

5.6.2 Hyperbolic cooling tower shell

A comprehensive buckling study of a concrete cooling tower shell under static wind loading is presented by Cole *et al.* in Ref. 7. In that exercise, a proprietary finite element program that has given very close comparisons to SHORE-III results for static and dynamic loading cases was used.

The shell geometry and vertical wind profile are shown in Fig. 5.1. The circumferential wind pressure distribution is similar to that shown in Fig. 3.19, and the stresses due to wind pressure along a selected meridian (θ = constant) are taken for the assumed symmetric prebuckled stress state $[\check{N}_i^j]$. This meridian could generally correspond to either $\theta = 0°$, where N_θ is a maximum in compression while N_ϕ is a maximum in tension; or, to $70° < \theta < 75°$ (Fig. 3.19), where N_ϕ may be maximum in compression. Meridian $\theta = 0°$ was reported as critical for all cases described in Ref. 8. The computed buckling pressures may be compared to the pressure at the top of the tower to compute a buckling factor of safety, which, in reality, is just the minimum eigenvalue $\bar{\lambda}^j$.

In Table 5.1, some results are summarized. One important modeling refinement illustrated is the treatment of the boundary conditions. Considered are, (H) an idealized situation very close to the hinged case described in Table 2.3; and (F) a more realistic flexible case which approximates the supporting columns (Section 3.3.5). For the circumferential wind pressure variation, one case of complete axisymmetry is included while, for the vertical variation, both the distribution shown in Fig. 5.1 and a constant value equal to the pressure at the top are studied. Also, constant and as-built thickness variations are considered along with top (cornice) and bottom (lintel) stiffening rings. Reported are the critical

Finite element analysis of shells of revolution

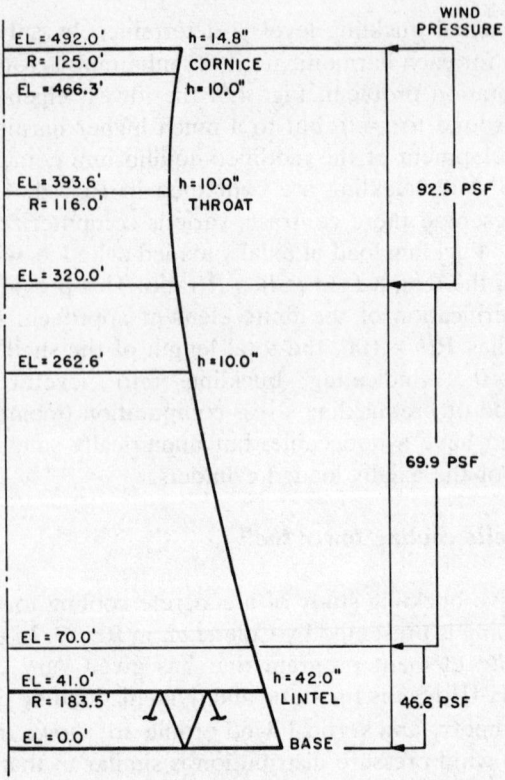

Fig. 5.1 Trojan cooling tower geometry

pressure values and the circumferential harmonic for the lowest eigenvalue. On Fig. 5.2, the mode shapes for several of the cases are also shown. No gravity loading was included in Cases 1–9.

Comparing Cases 1, 2, 3, the authors note that the use of an axisymmetric, constant pressure is very conservative. A modest reduction in buckling capacity can be attributed to the flexible support by comparing Cases 3 and 4. In Cases 4, 5, 6, the importance of considering the actual shell thicknesses is demonstrated. The modest increase in critical pressure achieved by only thickening the cornice region (Case 4 versus Case 5) is found noteworthy by the authors because of the higher circumferential wave-number and the change in mode shape, Fig. 5.2. This effect is also examined in the case study in Chapter 7.

Case 6 is said to be the most representative of the prototype and, by comparison with Cases 7 and 8, it is shown that the tower stiffened by bottom thickening is less sensitive to the vertical wind

Table 5.1

Case	Support	Pressure variation Circum.	Vertical	Thickness (in.)	Critical pressure (psf)	Harmonic j
1	H	Asym	Const.	Const. 10 in	630	5
2	H	Nsym	Const.	Const. 10 in	840	5
3	H	Nsym	Step	Const. 10 in	1000	6
4	F	Nsym	Step	Const. 10 in	880	5
5	F	Nsym	Step	Const. 10 in Cornice Fig. 5.1	920	8
6	F	Nsym	Step	Fig. 5.1	1880	7
7	F	Nsym	Const.	Fig. 5.1	1710	6
8	H	Nsym	Step	Fig. 5.1	1800	7
9	F	Nsym	Step	Const. 10 in + cornice and lintel from Fig. 5.1	1670	7
6S	F	Nsym	Step	Fig. 5.1	1840	7
6ST	F	Nsym	Step	Fig. 5.1	1845	7

gradient and to the support conditions than the constant thickness versions, Cases 2, 3 and 4. Cases 9 and 6 indicate that benefits beyond that provided by the top and base rings are realized by considering the actual thickness profile, shown in Fig. 5.1.

An additional case shown is 6S, where the self-weight stresses due to concrete with a density of 150 pcf are superimposed on Case 6. This results in a very slight decrease in critical pressure. Adding a 50 °F temperature gradient as well, Case 6ST, has little effect.

While the preceding example appears to be specialized, it illustrates several points which are felt to be somewhat more general. First, the most sensitive parameter which affects the stability of a shell with a given geometry and loading is the *thickness profile*. Also, realistic modeling of the *boundaries* is fairly important. Additionally, considerable benefits are realized by *circumferential stiffening* of the shell boundaries, which may be achieved by gradual thickening.

5.6.3 *Effect of imperfections*

One influence which is not addressed in this example is the effect of deviations from the ideal geometry on the buckling capacity of rotational shells. To study this effect properly, case studies for the shells of interest using realistic, as-built data should be considered, since construction

Finite element analysis of shells of revolution

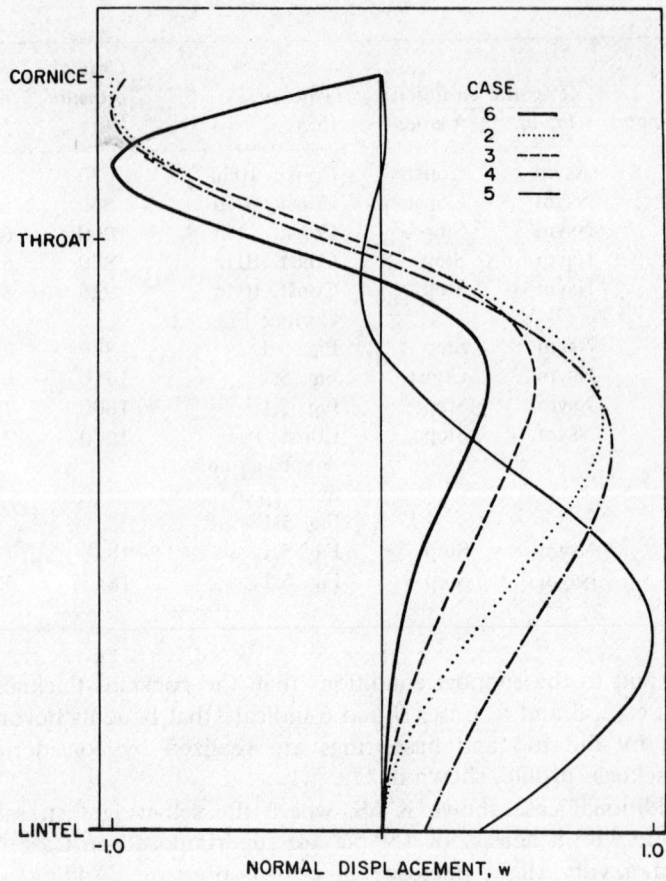

Fig. 5.2 Buckling mode shapes, simplified Trojan tower

and/or fabrication techniques would be the most likely origin of the deviations. While it is common to find stated in the literature that shells of negative Gaussian curvature are generally less 'imperfection sensitive' for instability analysis than those of zero or positive Gaussian curvature, there is considerable potential for utilizing finite element modelling, such as that described in this book, to quantitatively evaluate the buckling capacity of shells based on the *in situ* geometry.

With respect to the evaluation of the buckling capacity of shells with measured imperfections, it is suggested that the effect of such imperfections should not be over-exaggerated because of mathematical curve fitting to match the measured (surface) profile precisely. Abrupt fluctuations in the slope and change in curvature can generate correspondingly severe stress resultant and stress couple adjustments. It may be recalled

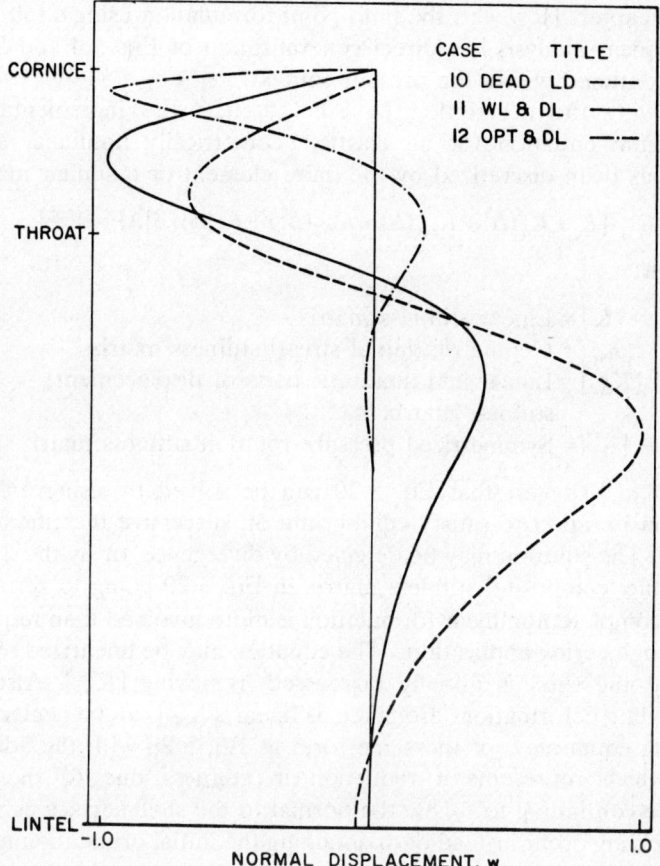

Fig. 5.2 (*contd.*) Buckling mode shapes, as-built Trojan tower

that shell theory is based on middle surface geometry which, particularly for thicker shells such as those constructed from reinforced concrete, may be considerably smoother than the surface variations upon which deviated geometry would usually be measured. Relatively low order least squares fits have been found to be useful in realistically incorporating measured deviations into the analysis.

5.7 Incremental nonlinear analysis

Beyond the determination of the critical or buckling load using Eq. 5.28, limit point;[9] post-buckling;[10] plastic buckling;[11,12] and ultimate strength[13] analyses may be of interest. The latter three topics are beyond the scope of this chapter. However, the limit point formulation using a full geometric nonlinear analysis is a direct generalization of Eqs 5.4 and 5.28 and can be outlined within the present context.

Following the description by Abel et al.,[9] the incremental global equilibrium equation for an elastic, geometrically nonlinear structure which has been discretized by the finite element or a similar method is

$$[\bar{K} + K_\sigma(\Delta) + K_{\Delta 1}(\Delta) + K_{\Delta 2}(\Delta^2) + K_f(\Delta)] \, \mathrm{d}\{\Delta\} = \mathrm{d}\{\bar{\mathcal{F}}\} \tag{5.29}$$

in which

$[\bar{K}]$ = Linear stiffness matrix
$[K_\sigma]$ = Geometric (initial-stress) stiffness matrix
$[K_{\Delta 1}], [K_{\Delta 2}]$ = Linear and quadratic parts of displacement stiffness matrix
$[K_f]$ = Symmetrized pressure-rotation stiffness matrix

Abel et al.[9] suggest that Eq. 5.29 can be solved by using a Newton–Raphson iteration to satisfy equilibrium on successive increments of the loading. The solution may be detected by divergence, or by the determinant of the 'composite' stiffness matrix in Eq. 5.29 going to zero.

The complete nonlinear formulation is more involved than required for many engineering applications. The equation may be linearized by assuming that the shell is initially unstressed, removing $[K_{\Delta 2}]$. Also, if the prebuckling deformations are taken as linear, $[K_{\Delta 1}]$ can be neglected. The resulting equation is of the same form as Eq. 5.28 with the addition of $[K_f]$, which represents a reduction in stiffness due to the applied pressure continuing to *follow* the normal to the shell surface as rotations and buckling occur instead of maintaining the initial orientation normal to the undeformed surface.

The pressure–rotation stiffness term is important for the analysis of axially compressed circular rings,[9] but seems to be less so for shells which are subjected to rapidly varying circumferential loading, such as wind-loaded hyperboloids, Fig. 3.19. However, it does not seem to be fully justified to neglect this term outright, as we did in writing Eq. 5.28 and as is done in some current computer programs. Thus, the more appropriate form of the classical buckling equation is

$$([\bar{K}^i] + \bar{\lambda}^i([K_\sigma^i] + [K_f^i])) \, \mathrm{d}\{\Delta^i\} = 0 \tag{5.30}$$

A comparison of the results of a full nonlinear analysis, using Eq. 5.29, to those obtained using a classical buckling analysis is of interest. Abel et al. reanalyzed the Trojan tower shell shown in Fig. 5.1 using a full two-dimensional discretization with general shell elements, similar to those described in Section 6.2. The wind loading included a suction pressure which increased the normalized Fourier coefficient for harmonic

$j = 0$ by 0.5 (See inset Fig. 3.19 where the value on the windward meridian is shown as *1.524* times the stagnation pressure p, the 0.524 being the suction coefficient) and computed a critical pressure of 538 psf. For a classical analysis, Eq. 5.30, the pressure was about 665 psf while the value for an equivalent axisymmetric loading was about 415 psi. The latter value, computed from the 2-D program, may be compared to Case 1, Table 5.1, obtained from an axisymmetric finite element program, if the latter value is adjusted to include the suction pressure, i.e., $630 \div 1.5 = 420$ psf.

While definitive conclusions should not be drawn on the basis of one example, it may be appropriate to employ a full nonlinear analysis for important or unusual cases.

References

1. Novozhilov, V. V., *Foundations of the Nonlinear Theory of Elasticity*, Graylock Press, Rochester, NY, 1953.
2. Stricklin, J. A., 'Geometrically Non-linear Static and Dynamic Analysis of Shells of Revolution', *Proc. of the Symposium of IAUTM in High Speed Computing of Elastic Structures*, University of Liege, 1979, pp. 383–411.
3. Klein, S., 'The Nonlinear Dynamic Analysis of Shells of Revolution with Asymmetric Properties by the Finite Element Method', Dissertation, University of Southern California, June, 1971, pp. 156–157.
4. Navaratna, D. R., Pian, T. H. H. and Witmer, E. A., 'Stability Analysis of Shells of Revolution by the Finite Element Method,' *AIAA J.*, Vol. 6, No. 2, 1968, pp. 355–361.
5. Zienkiewicz, O. C., *The Finite Element Method in Engineering Science*, McGraw-Hill, New York, 1971, pp. 414–425.
6. Gould, P. L. and Basu, P. K., 'Geometric Stiffness Matrices for the Finite Element Analysis of Rotational Shells', *J. Struct. Mech.*, Vol. 5, No. 1, 1977, pp. 87–105.
7. Cole, P. B., Abel, J. F. and Billington, D. P., 'Buckling of Cooling Tower Shells: Bifurcation Results', *J. Struct. Div. ASCE*, Vol. 101, No. ST6, June 1975, pp. 1205–1222.
8. Bushnell, D., 'Computerized Analysis of Shells-Governing Equations', *J. Computers and Structures*, Vol. 18, No. 3, 1984, pp. 47–51, pp. 471–536.
9. Abel, J. F., Chang, S.-C. and Hanna, S. L., 'Comparison of Complete and Simplified Elastic Buckling Loads for Cooling Tower Shells', *Proc. 2nd Int. Symp. on Natural-Draft Cooling Towers*, IASS, Ruhr-University Bochum, Springer-Verlag, Berlin, 1984.
10. Seide, P., *A Reexamination of Koiter's Theory of Initial Postbuckling Behavior and Imperfection Sensitivity in Thin Shell Structures* (edited by Y. L. Fung and E. E. Sechler), Prentice-Hall Inc., Englewood Cliffs, NJ, 1974, pp. 59–80.
11. Bushnell, D., *Plastic Buckling, Pressure Vessels and Piping: Design*

Technology—1982, A Decade of Progress (edited by S. Y. Zamirik and D. Dietrich), ASME, New York, 1982, pp. 47–117.
12. Wunderlich, W., Rensch, H. T. and Obrecht, H., 'Analysis of Elastic-Plastic Buckling and Imperfection-Sensitivity of Shells of Revolution: Buckling of Shells' (edited by E. Ramm), *Proc. of a State-of-the-Art Colloquium*, Springer-Verlag, Berlin, 1982, pp. 137–174.
13. Mang, H. A., Floegl, H., Trappel, F. and Walter, H., 'Wind-Loaded Reinforced-Concrete Cooling Towers: Buckling or Ultimate Load', *Engineering Structures*, Vol. 5, 1983, pp. 163–180.

6 Analysis of locally non-axisymmetric shells

6.1 Concept

Many shells of revolution encountered in industrial applications have local irregularities which alter the basic axisymmetry of the shell. Examples are cut-outs, pipe connections, and construction imperfections. Once a deviation is created on a rotational shell, the structure is no longer technically a shell of revolution, at least in the region of the deviation. Thus, the techniques presented in the earlier chapters are apparently not strictly applicable. An obvious recourse is to employ general type (triangular or quadrilateral) elements to model the entire shell. This requires a two-dimensional discretization, circumferential as well as meridional, and is demanding of computational resources, as compared to the basically one-dimensional rotational shell models.

In this chapter, a more concise technique is developed for the static analysis of locally non-axisymmetric shells. This model combines three different shell elements in a single analysis: rotational, general and transitional. The rotational elements are used where the shell is axisymmetric, while the general elements are specified over a limited region enveloping the irregularity, as shown in Fig. 6.1(a). The transitional elements facilitate a smooth transition between the two domains. Since the rotational shell elements are computationally more efficient than the general shell elements, this concept takes advantage of the prevailing axisymmetry, and at the same time provides the required generality for the non-axisymmetric portion.

The insertion of transitional elements is necessary to achieve continuity of displacement fields. As developed in the preceding chapters, the rotational element shown in Fig. 6.1(b) has *nodal circles*, and uses *trigonometric* Fourier series' to express displacement fields in the circumferential direction; whereas, the general element shown in Fig. 6.1(c) has *nodal points* and uses *polynomial* shape functions between the nodes. Since the two are not compatible, the displacement fields will be discontinuous if the elements are adjacent. A transitional element provides the required continuity between the two elements by employing a *line node*

Finite element analysis of shells of revolution

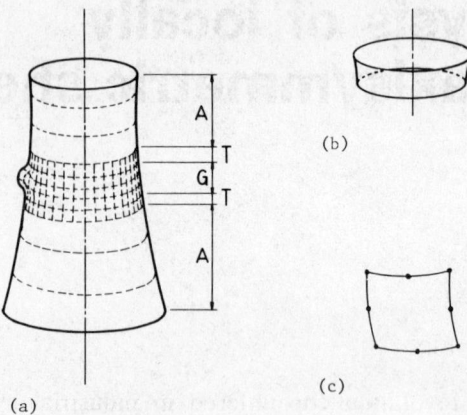

Fig. 6.1 FE mesh using different elements. (a) Combined mesh; (b) rotational shell element; (c) general shell element

and *point nodes* within a single element, as indicated in Fig. 6.2. The details of the formulation for the general and transitional elements are somewhat beyond the scope of this book, but are available elsewhere.[1-3] Only the basic definitions of geometry and displacements which are essential for the consideration of compatibility will be discussed.

The form of the global stiffness matrix for this model is worthy of special consideration. Non-zero terms are *not* limited to a narrow band along the diagonal, but are diffused throughout the matrix. This is due to the following: (1) two different types of nodes, ring nodes and point nodes, are combined in a single analysis; and (2) locally non-axisymmetric geometry creates a coupling between the harmonic coefficients of the rotational elements. Yet, the matrix still contains many zero terms that are scattered, but which should be considered for numerical efficiency. A

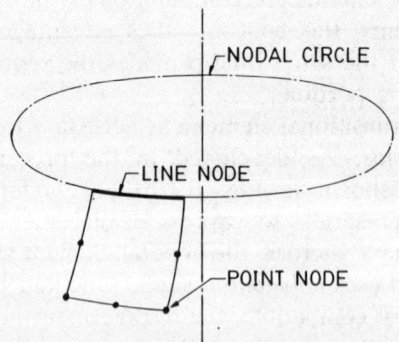

Fig. 6.2 Line and point nodes

solution algorithm that is effective for this situation is suggested and some numerical examples are presented.

6.2 General shell element

This element is patterned after the super-parametric element developed by Ahmad et al.,[1] incorporating the reduced integration technique of Zienkiewicz et al.[4] The only modification is the use of *cylindrical* coordinates (R, θ, Z) for the description of the geometry (Fig. 6.3), instead of *rectangular* Cartesian coordinates $(X, Y$ and $Z)$ which are employed in the 'Ahmad' element. Cylindrical coordinates are preferable for this application in that they provide a better representation of the basically axisymmetric geometry of the shell; and, they facilitate a simpler formulation when this element is combined with the rotational element in a single analysis.

To define the geometry, two vectors must be specified at each node i on the middle surface, as shown in Fig. 6.3. The first is the position vector

$$\mathbf{p}_i = \{R_i \ \theta_i \ Z_i\} \tag{6.1}$$

where R_i, θ_i and Z_i are the components in cylindrical coordinates, as defined in Section 2.1, and the second is a normal vector

$$\bar{\mathbf{n}}_i = \{\bar{R}_i \ \bar{\theta}_i \ \bar{Z}_i\} \tag{6.2}$$

with *magnitude* equal to the shell thickness, h_i, at the point, and with corresponding cylindrical components $\bar{R}_i, \bar{\theta}_i, \bar{Z}_i$. Next, local curvilinear coordinates ξ, η and ζ, ranging between $+1$ and -1, are established for each element, and the cylindrical geometry is equated as

$$\begin{Bmatrix} R \\ \theta \\ Z \end{Bmatrix} = \sum_i N_i(\xi, \eta) \begin{Bmatrix} R_i \\ \theta_i \\ Z_i \end{Bmatrix} + \sum_i N_i(\xi, \eta) \frac{\zeta}{2} \begin{Bmatrix} \bar{R}_i \\ \bar{\theta}_i \\ \bar{Z}_i \end{Bmatrix} \tag{6.3}$$

where $N_i(\xi, \eta)$ is a shape function used for interpolation. Details of this function are provided in Ref. 3.

Strains and stresses are defined with assumptions identical to those enforced for the rotational shell element, i.e. normal stresses are neglected and transverse shearing strains are included. Thus, this element also has five degrees of freedom at each node, three *Cartesian* translational components $\{U_i \ V_i \ W_i\}$ and two rotational components $\{B_{\phi i} \ B_{\theta i}\}$ about the positive circumferential and the negative meridional directions of the shell, respectively, as shown in Fig. 6.3. Also, as indicated on the figure,

Finite element analysis of shells of revolution

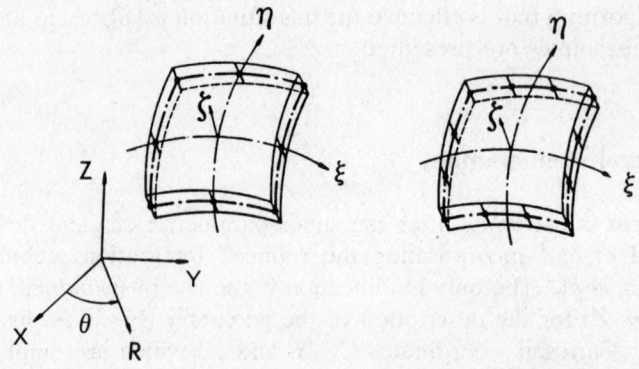

(a) Cylindrical Coordinates and local Geometry

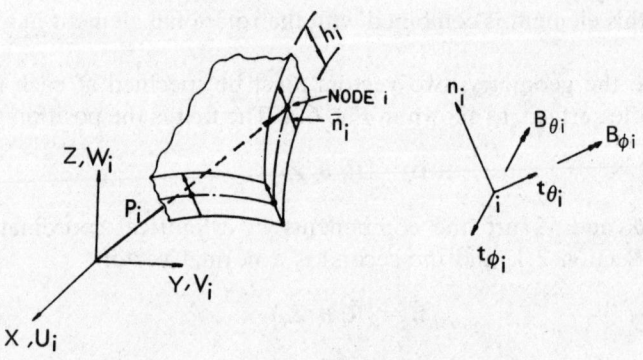

(b) Global Geometry

Fig. 6.3 General shell element. (a) Cylindrical coordinates and local geometry; (b) global geometry

unit vectors

$$\mathbf{t}_{\phi_i} = \{t_{\phi_i X} \, t_{\phi_i Y} \, t_{\phi_i Z}\} \tag{6.4a}$$

$$\mathbf{t}_{\theta_i} = \{t_{\theta_i X} \, t_{\theta_i Y} \, t_{\theta_i Z}\} \tag{6.4b}$$

$$\mathbf{n}_i = \mathbf{t}_{\phi i} \times \mathbf{t}_{\theta i} = \frac{\bar{\mathbf{n}}_i}{h_i} \tag{6.4c}$$

with corresponding Cartesian components are erected parallel to the positive meridional and circumferential directions, respectively. Then, the displacements may be expressed in the global Cartesian coordinates as

$$\begin{Bmatrix} U \\ V \\ W \end{Bmatrix} = \sum_i N_i \begin{Bmatrix} U_i \\ V_i \\ W_i \end{Bmatrix} + \sum_i N_i \zeta \frac{h_i}{2} \begin{bmatrix} t_{\phi_i X} & t_{\theta_i X} \\ t_{\phi_i Y} & t_{\theta_i Y} \\ t_{\phi_i Z} & t_{\theta_i Z} \end{bmatrix} \begin{Bmatrix} B_{\phi i} \\ B_{\theta i} \end{Bmatrix} \tag{6.5}$$

6.3 Transitional element

This element borders a rotational element on one side, a general element on the opposite side, and joins with other transitional elements on the remaining two sides. In this mesh arrangement, the transitional element must provide continuity of displacement fields, as well as geometry, with adjoining elements. The transitional element is formed from the general shell element described in the preceding section. Compatibility would be achieved automatically on three sides of the element, but not on the boundary with the rotational element. If a modification is then made such that the element is also compatible with the rotational element on the remaining boundary, then all of the continuity requirements would be fulfilled. The transitional element achieves this by adopting a *line node* which extends all along the common boundary with the rotational element and which can accommodate any function along the line. The line node, shown on Fig. 6.4, possesses three sub-nodes, one at each end of the line and a *moving node* which may be located anywhere on the line. As a set, these three nodes can represent any function that the adjoining element may impose on the common boundary. Details of the development of the line node, including the shape function, are given in Ref. 3. On the remaining three sides, the transitional element has the same point nodes, with the identical shape functions, as those of the general element.

Once the shape functions are developed, the remaining formulation follows similar steps as the general shell element. Thus, the position and normal vectors are specified at each nodal point, including the moving node; the geometry is defined by Eq. 6.3; and, the displacements are expressed by Eq. 6.4. However, there are some differences in the application of these equations. First, it should be noted that the coordinate θ_i for defining the moving node in Eq. 6.3 is a variable instead of a constant,

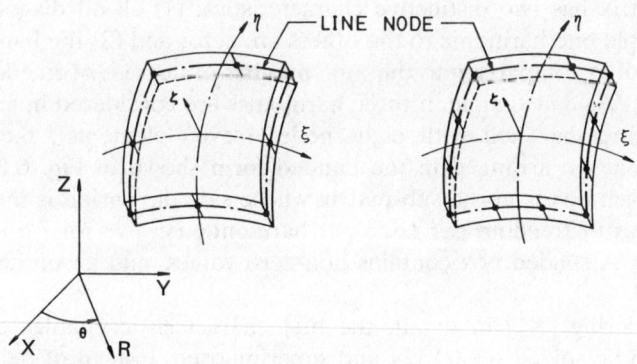

Fig. 6.4 Transitional shell element

since the node itself may move along the boundary. Second, Cartesian components $\{U\,V\,W\}$ are used for the transitional and general elements, while curvilinear displacement components $\{u\,v\,w\}$ are used for the rotational element. Thus, a coordinate transformation is required along the line node. Furthermore, along each nodal circle, the generalized displacements for the transitional element are physical displacement components, while those for the rotational element are Fourier coefficients. Thus, at each sub-node, the nodal degrees of freedom must be related to the harmonic degrees of freedom.

The resulting form of the displacement transformation at each sub-node i along the line node is

$$\begin{Bmatrix} U_i \\ V_i \\ W_i \end{Bmatrix} = [T_i] \sum_j \begin{bmatrix} \cos j\theta_i & 0 & 0 \\ 0 & \sin j\theta_i & 0 \\ 0 & 0 & \cos j\theta_i \end{bmatrix} \begin{Bmatrix} u^j \\ v^j \\ w^j \end{Bmatrix} \quad (6.6)$$

and

$$\begin{Bmatrix} B_{\phi i} \\ B_{\theta i} \end{Bmatrix} = \sum_j \begin{bmatrix} \cos j\theta_i & 0 \\ 0 & \sin j\theta_i \end{bmatrix} \begin{Bmatrix} \beta_\phi^j \\ \beta_\theta^j \end{Bmatrix} \quad (6.7)$$

where the coordinate transformation matrix is[3]

$$[T_i] = [\mathbf{t}_{\phi_i}\ \mathbf{t}_{\theta_i}\ \mathbf{n}_i] = \begin{bmatrix} t_{\phi_i X} & t_{\theta_i X} & \bar{n}_{iX}/h_i \\ t_{\phi_i Y} & t_{\theta_i Y} & \bar{n}_{iY}/h_i \\ t_{\phi_i Z} & t_{\theta_i Z} & n_{iZ}/h_i \end{bmatrix} \quad (6.8)$$

6.4 Global system

The complete global stiffness matrix for a linear shell of revolution analyzed with only rotational elements would consist of matrices $[\bar{K}^j]$, Eq. 3.38, arranged along the diagonal for all participating harmonics. This matrix has two distinctive characteristics: (1) all off-diagonal terms that couple one harmonic to the others are zero; and (2) the bandwidth is very small in comparison to the total number of degrees of freedom of the structure. As an example, if three harmonics are considered in an analysis of a cylindrical shell with eight nodes (seven elements), the stiffness matrix may be arranged in the banded form shown in Fig. 6.5. In that figure, each box is also a sub-matrix whose side dimension is the number of degrees of freedom per node per harmonic, i.e. five for $j \geq 1$ or three for $j = 0$. A shaded box contains non-zero values, and an unshaded box, zeros.

Considering $[\bar{K}^j]$ in detail, the first characteristic enables each harmonic to be solved separately and superimposed, instead of being considered simultaneously, and makes it unnecessary to ever assemble the

Analysis of locally non-axisymmetric shells

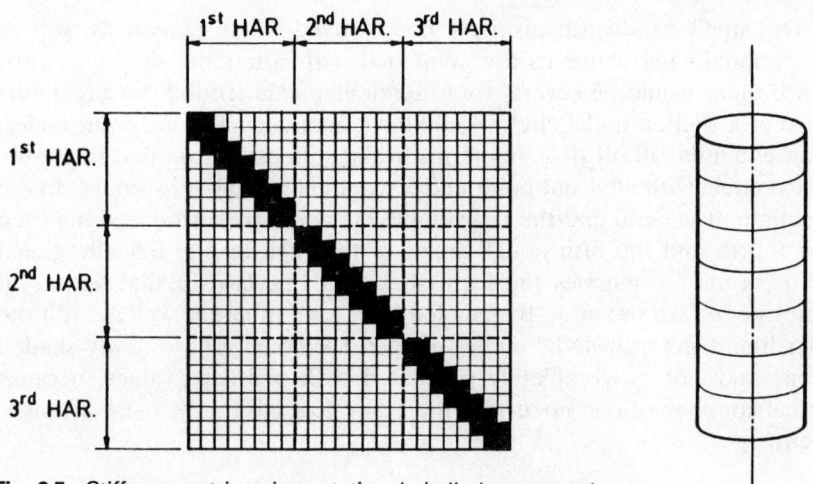

Fig. 6.5 Stiffness matrix using rotational shell elements only

entire matrix for the purely axisymmetric geometry linear case. The second characteristic is attributable to the fact that a rotational element has nodal circles, instead of nodal points. Since any internal node has only two neighboring nodes (except in the case of a branched shell), the semi-bandwidth includes the nodal degrees of freedom of only two nodes, Fig. 6.5.

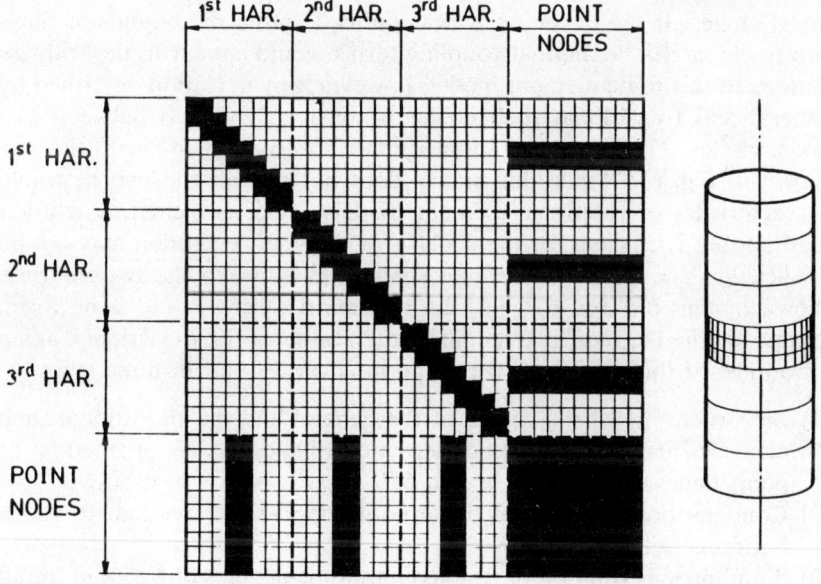

Fig. 6.6 Stiffness matrix using rotational, general and transitional shell elements

153

Finite element analysis of shells of revolution

The small bandwidth property is destroyed when general as well as transitional shell elements are combined with rotational shell elements. Since there would be several transitional elements around the circumference of a shell, a nodal circle would have many neighboring point nodes. Consequently, all off-diagonal terms in the stiffness matrix that relate the nodal circle with adjacent point nodes would have non-zero values. In the example, if general and the transitional elements are introduced between the fourth and the fifth nodal circles of the shell in Fig. 6.5, the global stiffness matrix assumes the form shown in Fig. 6.6. In that figure, all point nodes are shown to be coupled with each other, as well as with the fourth and the fifth nodal circles in each harmonic. These newly shaded areas may not be completely populated with non-zero values, because actual coupling of one node with the other depends on the mesh arrangement.

6.5 Solution procedure

When a shell is non-axisymmetric, the off-diagonal terms that relate one harmonic with the others in the complete global stiffness matrix may no longer be null. For example, if the shell in Fig. 6.6 has non-axisymmetric geometry between the fourth and the fifth nodal circles, the form of the resulting stiffness matrix is as shown in Fig. 6.7. In this example, all harmonics are coupled through the fourth and the fifth nodal circles. If the deviation is not confined within the bounds of these two nodal circles, additional coupling terms would appear in the stiffness matrix. In this finite element model, however, any deviation described by general and transitional shell elements must be confined between two nodal circles. Therefore, the stiffness matrix would be similar to the one shown in Fig. 6.7. Since this matrix does not possess the two desirable characteristics of rotational element models previously cited, a solution method that is efficient for the analysis of shells of revolution may not be applicable. Yet, there are definite similarities between the two matrices shown in Figs 6.5 and 6.7, i.e. the two matrices are in the same form, except for the degrees of freedom in the region of the deviation. Taking advantage of this similarity, the solution is carried out in three steps:

(1) Substructuring of the region of the general and the transitional shell elements and subsequent condensation of the degrees of freedom of point nodes, marked 'P' in Fig. 6.7.
(2) Condensation of uncoupled harmonic degrees of freedom for each harmonic, marked 'UH'.
(3) Simultaneous solution of coupled harmonic degrees of freedom for all harmonics, marked 'CH'.

Analysis of locally non-axisymmetric shells

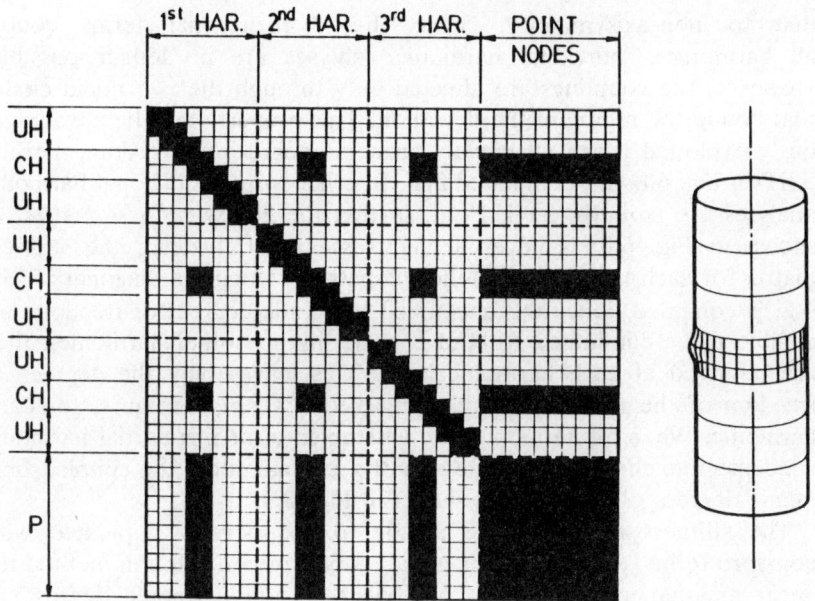

Fig. 6.7 Stiffness matrix of a shell of revolution with a local deviation

In the above procedure, the first and the second steps are interchangeable.

Although the presence of the point nodes creates several off-diagonal blocks that are filled with non-zero values in the global stiffness matrix, these blocks link the point nodes with only two nodal circles. Any remaining off-diagonal blocks that connect the point nodes with other nodal circles remain null. In order to avoid unnecessary numerical effort in dealing with many zero terms in the global stiffness matrix, the region between the two nodal circles is taken as a substructure. The stiffness matrix of the substructure consists of the portions of the global stiffness matrix that are marked 'P' and 'CH' in Fig. 6.7. The degrees of freedom of the point nodes marked 'P' in the figure are internal nodes of the substructure and may be eliminated by static condensation at the substructure level. The stiffness matrix of the substructure, after condensation, contains only the harmonic degrees of freedom of the two bounding nodal circles, marked 'CH' in the figure. Thus, the global stiffness matrix that is assembled with this reduced matrix for the substructure assumes a much simpler form, containing only harmonic degrees of freedom, as shown in Fig. 6.8.

Considering the stiffness matrix in the form of Fig. 6.8, it is narrowly banded along the diagonal, with non-zero terms scattered off the diagonal

Finite element analysis of shells of revolution

due to non-axisymmetry. Since these off-diagonal terms couple all harmonics, individual harmonic analyses are no longer possible. However, the couplings are affected only through the two nodal circles that bound the region of the deviation. This property may be advantageously exploited if the *uncoupled* harmonic degrees of freedom, marked 'UH' in Fig. 6.8, are eliminated first. In other words, individual harmonic analyses are *partially* carried out, so that the global stiffness matrix, as shown in Fig. 6.8, is never actually assembled. Instead, the stiffness matrix for each harmonic, i.e. the submatrices along the diagonal of Fig. 6.8, is compiled. Since the uncoupled harmonic degrees of freedom are eligible to be eliminated without considering any other harmonics, they are removed at each harmonic level. This leaves only the degrees of freedom of the two nodal circles, marked 'CH' in the figure, for each harmonic. When all harmonics are assembled after this partial harmonic condensation, the global stiffness matrix contains only the coupled harmonic degrees of freedom, as shown in Fig. 6.9.

The stiffness matrix shown in Fig. 6.9 is densely populated with non-zero terms, even off the diagonal. Any standard solution method for linear simultaneous equations will yield all the harmonic degrees of freedom of the two nodal circles, i.e. the Fourier coefficients of the displacements along the nodal circles. Once the harmonic degrees of

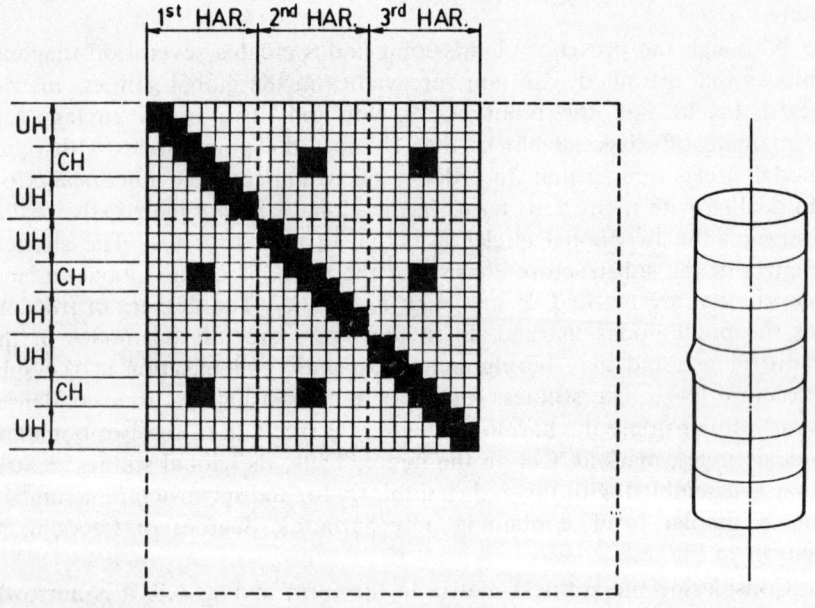

Fig. 6.8 Stiffness matrix after the substructuring

Analysis of locally non-axisymmetric shells

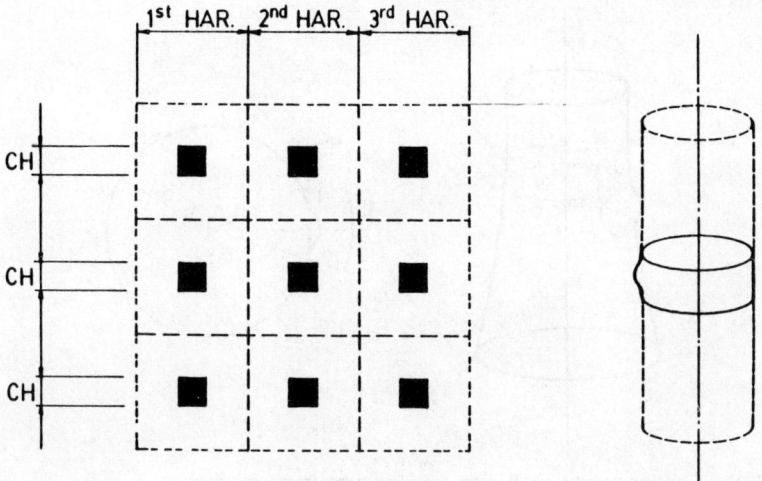

Fig. 6.9 Stiffness matrix after the separate harmonic analysis

freedom of the two nodal circles are found, the uncoupled harmonic degrees of freedom are computed by back-substitution, which may be carried out individually for each harmonic. Likewise, the back-substitution to find the displacements of the point nodes is carried out at the substructure level. This is a standard procedure that is similar to that discussed in the last paragraph of Section 3.2.3.3.

6.6 Case Studies

6.6.1 Imperfect hyperboloidal shell under self-weight and wind loading

A hyperbolic cooling tower shell with a bulge-type imperfection, shown in Fig. 6.10, is considered. The geometry of the perfect shell as well as the discretization for the rotational elements, which includes columns, are given in Fig. 6.11. The bulge is confined within the region marked 'Imperfection' and is horizontally centered on the zero degree meridian. The circumferential width of the bulge, 41.96 ft, is approximately the same as the vertical height, 41.60 ft, and the middle surface profile is expressed by

$$R = R_0 + \frac{A}{4}\left\{1+\cos\left[\frac{\pi(Z-Z_c)}{(l/2)}\right]\right\}\left\{1+\cos\left[\frac{\pi j(\theta-\theta_c)}{(\theta_\omega/2)}\right]\right\} \quad (6.9)$$

Finite element analysis of shells of revolution

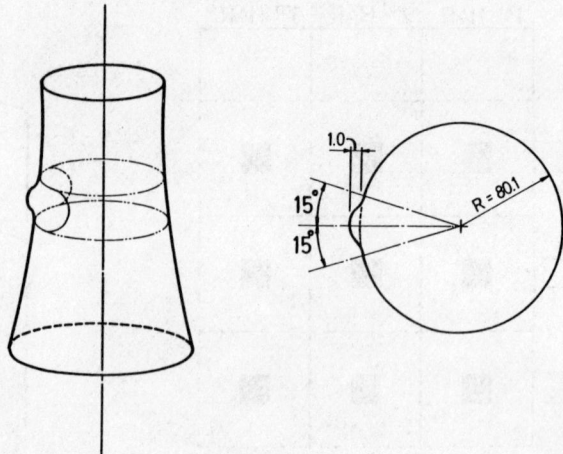

Fig. 6.10 Cooling tower shell with a bulge-type imperfection

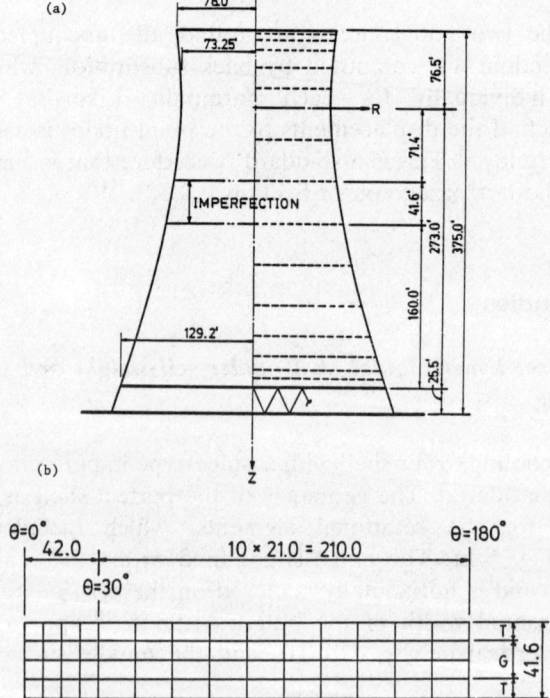

Fig. 6.11 Cooling tower shell. (a) Geometry and FE mesh for rotational elements; (b) FE mesh for the region of imperfection

where

$R_0(Z)$ = radius of the perfect shell.
A = the amplitude of the bulge at its center = +1 ft.
$Z_c = Z$ coordinate at the center of the bulge = 92.2 ft.
l = length of the bulge = 41.6 ft.
j = Fourier harmonic = 1.
$\theta_c = \theta$ coordinate at the center of the bulge = 0°.
θ_ω = the angle subtended by the bulge width = 30°.

General and transitional shell elements are used in the region marked 'Imperfection' to simulate the bulge geometry. The region is sub-discretized by four even rows meridionally and by fourteen variable intervals circumferentially, as shown in Fig. 6.11. The first four columns are at 7.5 degree and the last ten columns are at 15 degree increments. Since the geometry is symmetric about the zero degree meridian, only one-half of the circumference (from 0 to 180°) is used for the analysis of the substructure. The material properties are taken as follows:

Modulus of elasticity: E = 519 100 kips/sq. ft
Poisson's ratio: $\mu = 1/6$
Shear factor: $\lambda = 5/6$.

This shell is analyzed for two loading cases: self-weight and static wind. First, the results for the self-weight analysis are considered. In Figs 6.12, the stress resultants are plotted along the meridian at $\theta = 2°$, which is near the center of the bulge. In Fig. 6.12(a), values for the perfect shell are shown; while in Fig. 6.12(b), the corresponding values for the 'bulged' geometry are given. Also shown with dashed lines in the later figure are the results when an axisymmetric imperfection is assumed, i.e. $j = 0$ in Eq. 6.9. The latter form is frequently used as an approximation for a bulge, especially if only an axisymmetric analysis capability is available. In examining Fig. 6.12(b), it may be seen that the circumferential forces are considerably affected, even by an axisymmetric representation of the imperfection, but that the meridional forces are not. On the other hand, the actual bulge imperfection creates disturbances in both meridional and circumferential forces.

Since the shell with the bulge is no longer axisymmetric, it is of interest to examine the stress resultants around the circumference of the shell. These diagrams are plotted with solid lines near the mid-height of the bulge in Fig. 6.13. In addition to the results for the asymmetrically bulged shell, two other sets are shown for comparison: The stresses for the perfect shell; and, the stresses calculated assuming that the imperfection extends around the entire shell and is thus axisymmetric. It may be

Finite element analysis of shells of revolution

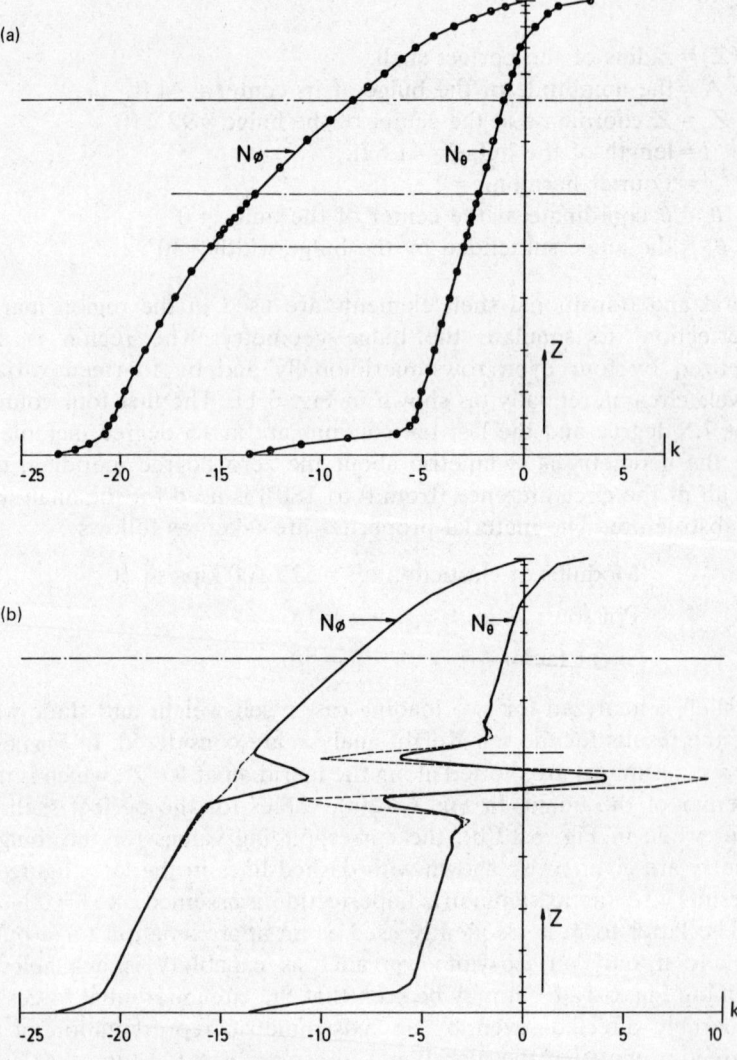

Fig. 6.12 (a) Stress resultants (dead load); (b) stress resultants (DL-outward bulge) $\theta = 2°$

noted that the disturbances of the stress resultants due to the bulge is limited to the vicinity of the imperfection; whereas, at points far from the bulge, the stress resultants for the shell with the bulge are the same as those of the perfect shell. Figures 6.12 and 6.13 also show that the bulge imperfection does not affect the circumferential forces as much as the

Analysis of locally non-axisymmetric shells

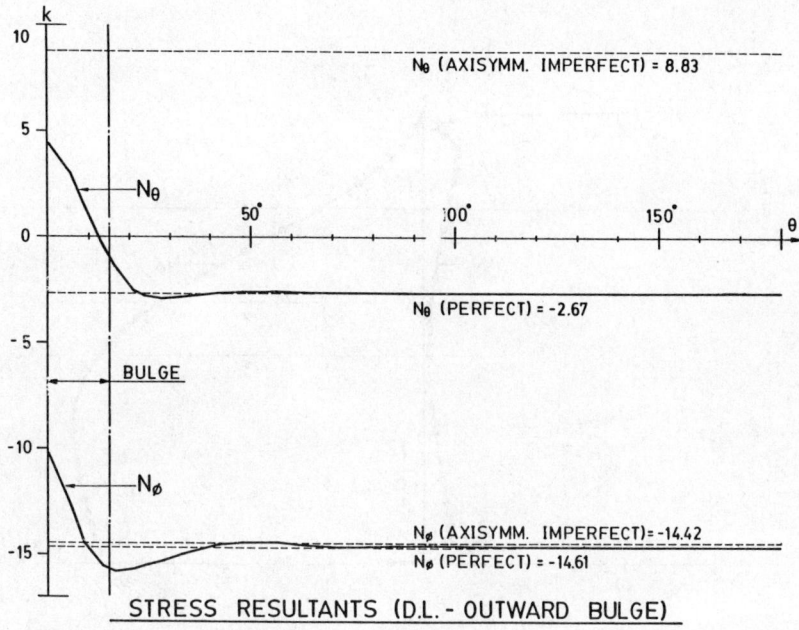

Fig. 6.13 Circumferential distribution of stress resultants—dead load, k

axisymmetric imperfection and that the meridional forces are more severely affected by the bulge.

Stress couples are not plotted. Even though stress couples are also affected by the imperfection, they are still small in comparison to the stress resultants.[5]

Next, static wind pressure loading similar to that shown on Fig. 3.19 and taken as constant over the height of the shell is considered. The circumferential distribution of the pressure is expressed in terms of the following six Fourier harmonic loading components:

Harmonic number, j	Fourier components, f_w^j
0	−0.003836
1	−0.009268
2	−0.021457
3	−0.017974
4	−0.003790
5	+0.003377

Stress resultants and couples along the $\theta = 2°$ meridian are plotted in Fig. 6.14. In Fig. 6.14(a) and (b) the values for the perfect shell are shown, while in Fig. 6.14(c) and (d) the corresponding values for the 'bulged'

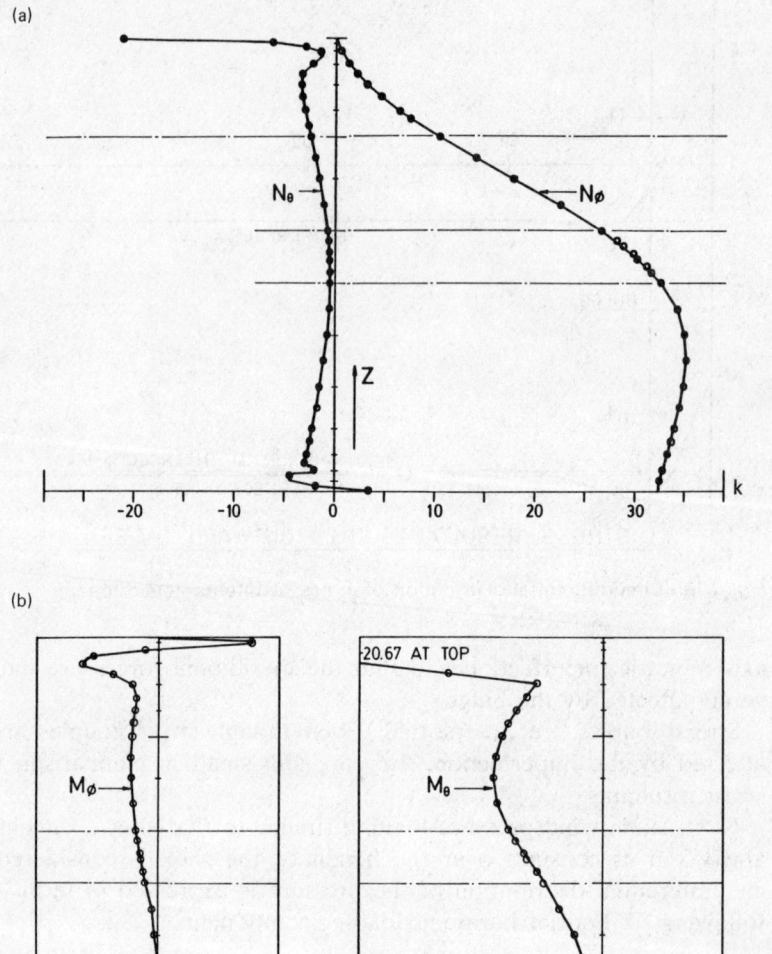

Fig. 6.14 Results of wind loading analysis. (a) Stress resultants (wl—perfect) $\theta = 2°$; (b) stress couples (wl—perfect) $\theta = 2°$

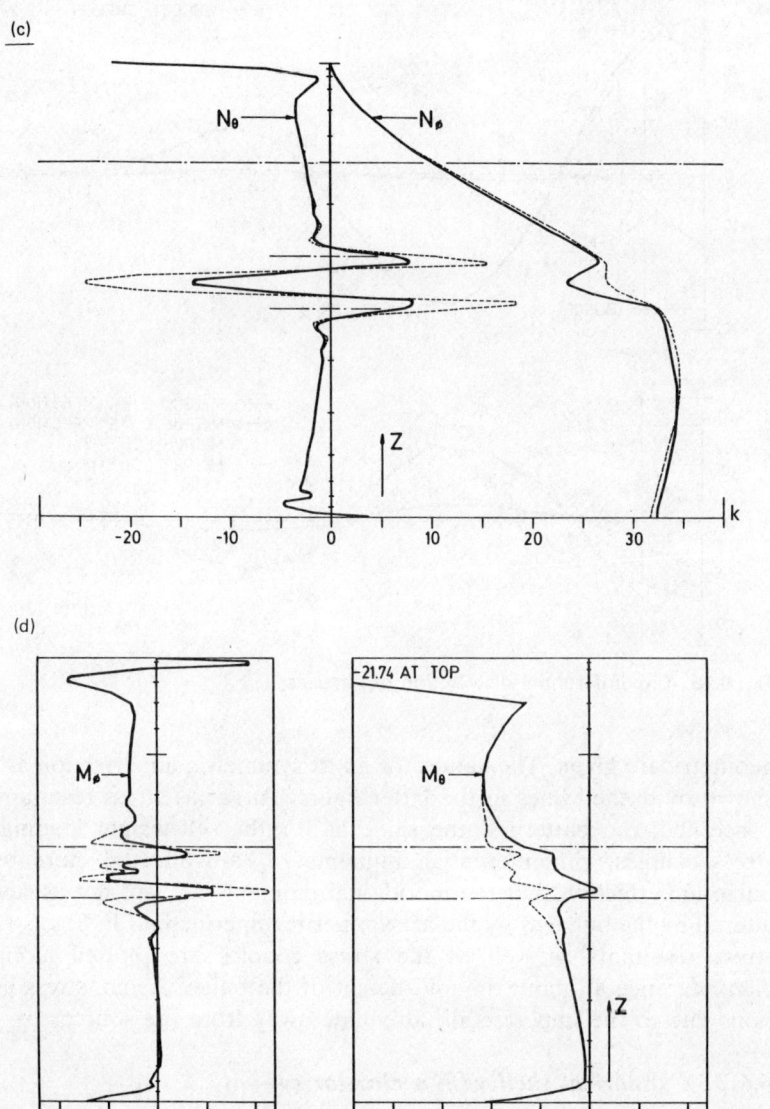

Fig. 6.14 (*contd.*) (c) Stress resultants (wl—outward bulge) $\theta = 2°$; (d) stress couples (wl—outward bulge) $\theta = 2°$

Finite element analysis of shells of revolution

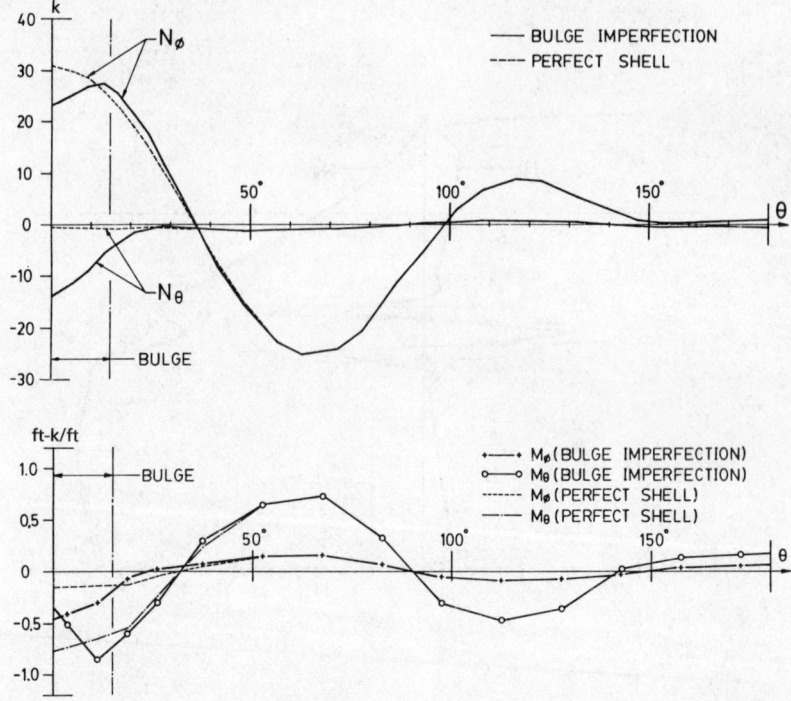

Fig. 6.15 Circumferential distribution of stresses

geometry are given. The values for an axisymmetric imperfection are also shown by dashed lines in the latter figure. As far as stress resultants are concerned, the pattern is the same as for the self-weight loading. For stress couples, circumferential moments M_θ are affected more by the bulge imperfection, but the meridional moments M_ϕ are not as severely altered by the bulge as by the axisymmetric imperfection. In Fig. 6.15, the stress resultants as well as the stress couples are plotted along the circumference at about the mid-height of the bulge. Again, stress alterations due to the imperfection attenuate away from the source.

6.6.2 Cylindrical shell with a circular cut-out

An infinitely long cylindrical shell with a circular cut-out was analyzed by Van Dyke using a shallow shell theory.[6] Key studied a long cylinder with a circular cut-out, shown in Fig. 6.16, under axial tension and compared the results with those of Van Dyke to demonstrate the capability of a then newly developed quadrilateral finite element.[7] In this example, the same shell studied by Key is treated using the mesh shown in Figs 6.16

Analysis of locally non-axisymmetric shells

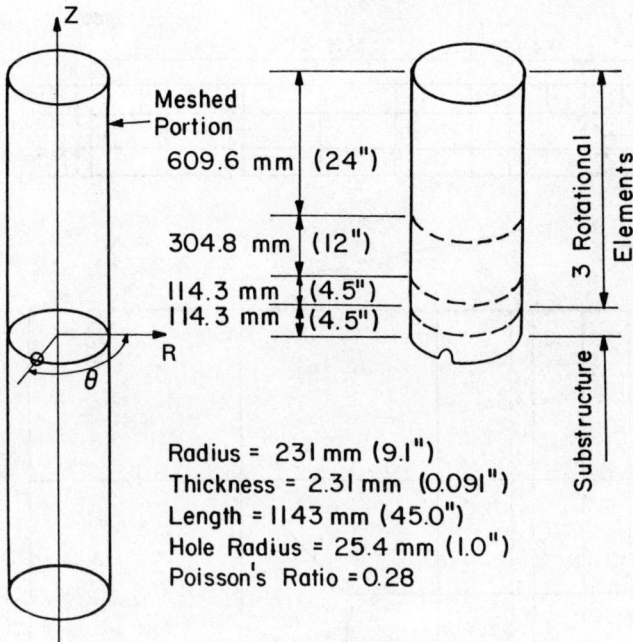

Fig. 6.16 Cylinder with a circular cut-out and the FE mesh

and 6.17. Taking advantage of the symmetry about $Z = 0$, only one-half of the length of the shell is considered, Fig. 6.16, while symmetry about $\theta = 0$ is utilized in the substructure mesh, Fig. 6.17.

The results of an analysis when 13 harmonics are employed are shown in Table 6.1, along with those of Van Dyke[6] and Key.[7] In Table 6.1, σ_m is the extensional stress; σ_b is the bending stress; σ_∞ is the membrane theory tensile stress at infinity; and points A and B are indicated on Fig. 6.17. It is seen that the locally non-axisymmetric finite element model requires only approximately *one-fourth* the degrees of freedom of the conventional finite element model to obtain comparable results. This translates into enormous savings in computing time when one considers that the CPU time required to solve simultaneous equations is nominally a cubic function of the number of equations.

Since the local non-axisymmetric deviation couples all Fourier harmonics, theoretically an *infinite* number of interdependent harmonic terms constitute the degrees of freedom of the rotational elements. This comparison focuses on the convergence as the total number of harmonics included in the analysis, $j = 0, 1, \ldots, \bar{j}$, is increased. Stress intensities (σ/σ_{00}) at points A and B (Fig. 6.17) are shown in Fig. 6.18. Though the

165

Finite element analysis of shells of revolution

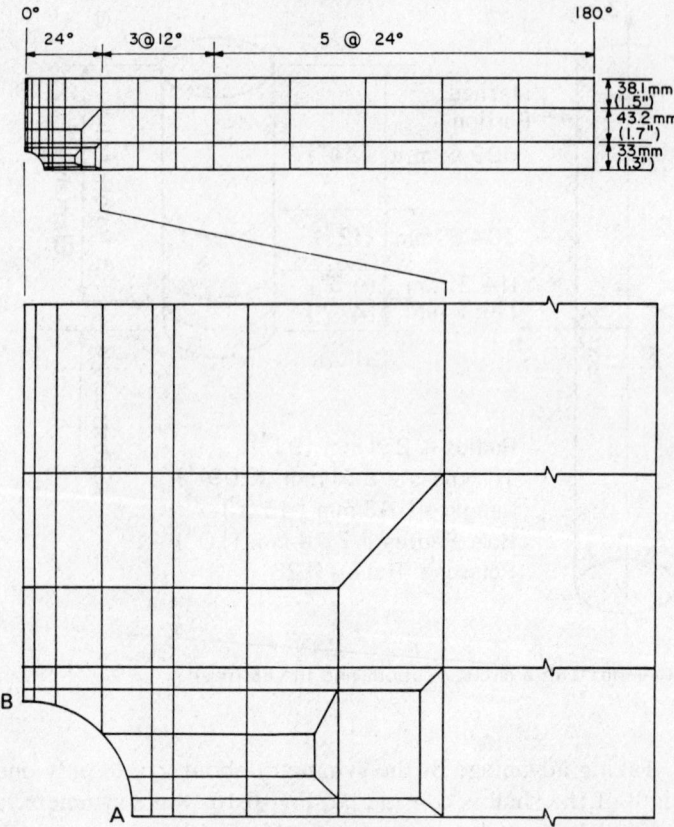

Fig. 6.17 Finite element mesh of the substructure

stresses change initially with increasing $\bar{j}$, the values stabilize fairly rapidly, e.g. when $\bar{j} = 12$ in this example. It is expected that a larger value of $\bar{j}$ would be required to represent the displacements of rotational elements if the length of the substructure were decreased, and vice versa. This is due to the fact that the effect of the geometric deviation is most severe in the immediate vicinity of the deviation.

6.7 Extensions

While the developments in this chapter indicate that the efficiency of modeling axisymmetrical shells with localized deviations without resort to a full general shell element model may be realized, further research is needed to exploit this technique for dynamic and stability problems, and for nonlinear situations. However, even for elastostatic problems, it is

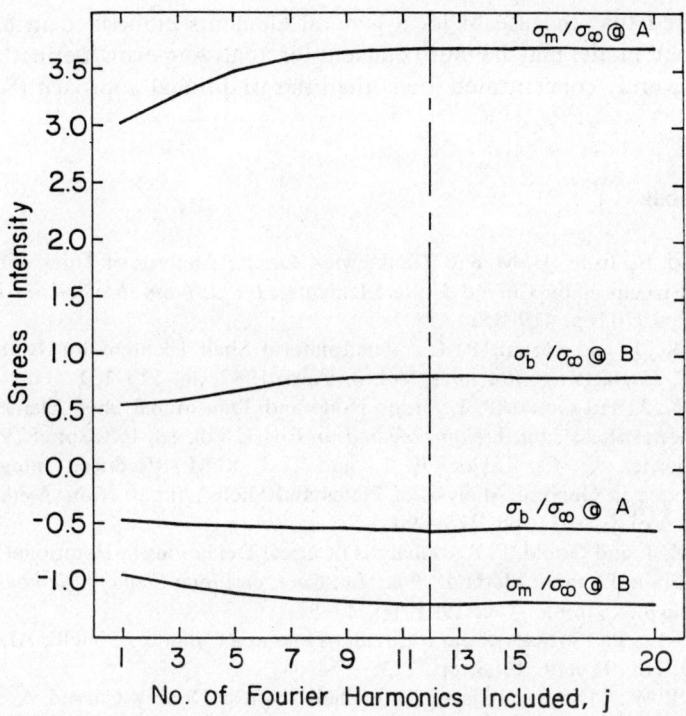

Fig. 6.18 Stress intensity around the hole. A is at the side of the hole at $R\theta = 25.4$, $Z = 0.0$; B is at the top of the hole at $R\theta = 0.0$, $Z = 25.4$

Table 6.1 Stresses for a cylinder under axial tension with a circular cut-out

	Van Dyke[6]	Key[7] Coarse mesh (1534 d.o.f.)	Key[7] Fine mesh (5301 d.o.f.)	Locally non-axisymmetric model (1447 d.o.f.)
σ_m/σ_∞ at A	3.60	3.44 (4.6%)	3.59 (0.3%)	3.66 (1.60%)
σ_b/σ_∞ at A	±0.59	±0.55 (7.4%)	±0.55 (7.4%)	±0.554 (6.06%)
σ_m/σ_∞ at B	−1.25	−1.08 (15.7%)	−1.18 (5.4%)	−1.20 (3.87%)
σ_b/σ_∞ at B	±0.809	±0.806 (0.4%)	±0.812 (0.4%)	±0.795 (1.71%)

A is at the side of the hole at $R\theta = 25.4$, $Z = 0.0$.
B is at the top of the hole at $R\theta = 0.0$, $Z = 25.4$.

anticipated that the use of local general elements embedded in a rotational shell model may be more efficient for analyzing even 'perfect' shells under severely concentrated loads than the traditional approach (Section 2.3.2).

References

1. Ahmad, S., Irons, B. M. and Zienkiewicz, O. C., 'Analysis of Thick and Thin Shell Structures by Curved Finite Elements', *Int. J. Num. Methods in Engrg*, Vol. 2, 1970, pp. 419–451.
2. Han, K. J. and Gould, P. L., 'Quadrilateral Shell Element for Rotational Shells', *Engineering Structures*, Vol. 4, April 1982, pp. 129–131.
3. Han, K. J. and Gould, P. L., 'Line Node and Transitional Shell Element for Rotational Shells', *Int. J. Num. Methods in Engrg*, Vol. 18, 1982, pp. 879–895.
4. Zienkiewicz, O. C., Taylor, R. L. and Too, J. M., 'Reduced Integration Technique in General Analysis of Plates and Shells', *Int. J. Num. Methods in Engrg*, Vol. 3, 1971, pp. 275–290.
5. Han, K. J. and Gould, P. L., 'Analysis of Local Deviations in Rotational Shells using Finite Element Method', *Proc. Int. Conf. on Finite Element Methods*, Vol. II, Shanghai, China, Aug. 1982, pp. 43–51.
6. Van Dyke, P., 'Stresses about a Circular Hole in a Cylindrical Shell', *AIAA J.*, Vol. 3, No. 9, Sept. 1965, pp. 1733–1742.
7. Key, S. W., 'The Analysis of Thin Shells with a Doubly Curved Arbitrary Quadrilateral Finite Element', *J. Computers and Structures*, Vol. 2, 1972, pp. 637–673.

7 Computer programs and case study

7.1 General

Numerical analysis of rotational shells has been carried out for many years with large scale computer codes. These programs have been based on a variety of mathematical formulations, such as numerical integration, transfer matrices, finite differences, and finite elements. Ultimately, even those which are not generically finite element based often assume the appearance of finite element modeling, with subdivided domains and discrete stations.

Many of the programs have remained proprietary, especially the enhanced versions. A fairly complete tabulation of first generation codes is contained in Ref. 1. In order to provide a flavor of the capabilities of some representative public domain programs, descriptions of the BOSOR4 and SHORE III programs are presented in the following section, without attempting to make a comparative assessment. Then, a relatively detailed case study is developed in which many of the capabilities of SHORE III are demonstrated.

7.2 Computer programs

7.2.1 BOSOR4[1]

The program performs stress, stability and vibration analysis of segmented, branched, ring-stiffened, elastic shells of revolution with various wall constructions. It is very general with respect to the meridian geometry, wall design, edge conditions, and loading. The capabilities of the program are summarized in Table 7.1.

The program performs three distinct types of analyses:

(1) A linear analysis for the behavior of axisymmetric shell systems under axisymmetric and non-symmetric loading.

Table 7.1 BOSOR4 capability summary[1]

Type of analysis	Shell geometry	Wall construction	Loading
Nonlinear axisymmetric stress	Multiple-segment shells, each segment with its own wall construction, geometry and loading	Monocoque, variable or constant thickness	Axisymmetric or non-symmetric thermal and/or mechanical line loads and moments
Linear symmetric or non-symmetric stress	Cylinder, cone, spherical, ogival, toroidal, ellipsoidal, etc.	Skew-stiffened shells	Axisymmetric or non-symmetric thermal and/or mechanical distributed loads
Stability with linear symmetric or non-symmetric prestress, or with nonlinear symmetric prestress	General meridional shape; point-by-point input	Fibre-wound shells	Proportional loading
Vibration with nonlinear pre-stress analysis	Axial and radial discontinuities in shell meridian	Layered orthotropic shells	Non-proportional loading
Variable mesh point spacing within each segment	Arbitrary choice of reference surface	Corrugated, with or without skin	
	General edge conditions	Layered orthotropic with temperature-dependent material properties	
	Branched shells	Any of above wall types reinforced by stringers and/or rings treated as 'smeared out'	
	Prismatic shells and composite, built-up panels	Any of above wall types further reinforced by rings treated as discrete	
		Wall properties variable along meridian	

(2) A nonlinear stress analysis for axisymmetric behavior of axisymmetrically loaded elastic shells.
(3) An eigenvalue analysis in which the eigenvalues represent buckling loads or vibration frequencies of axisymmetric shell systems under axisymmetric loading.

The analysis is based on energy minimization with constraint conditions. The total energy of the system includes the strain and kinetic energy of the shell segments and discrete rings, and the potential energy of the applied line and surface loads. The constraint conditions arise from prescribed displacements at the boundaries of the structure; anywhere within the structure; and at junctures between segments, and are introduced into the energy functional by means of Lagrange multipliers.

The linear stress analysis is carried out in harmonic form using a finite difference technique, which is energy-based. In the nonlinear axisymmetric stress analysis, the energy expression contains up to quadratic terms in the meridional and normal displacements. Energy minimization leads to a set of nonlinear algebraic equations that are solved by the Newton–Raphson method. Stress resultants and couples are calculated at the mesh points using the constitutive equations and the kinematic relations.

The results from the nonlinear axisymmetric or linear non-symmetric stress analysis are used in the eigenvalue analyses for buckling and vibration. The prebuckling meridional and circumferential stress resultants, N_ϕ and N_θ, and the meridional rotation, β_ϕ, appear as known variable coefficients in the energy expressions that govern buckling and vibration. These expressions are homogeneous quadratic forms that are rendered stationary with respect to infinitesimal variations of the dependent variables. The eigenvalues which are formed from the solution of the resulting algebraic equations then represent buckling loads or natural frequencies.

Further capabilities and applications of the BOSOR4 code are well documented in Ref. 1 and several companion papers cited in that publication.

7.2.2 SHORE III[2]

The program is designed for the linear static and dynamic analysis of shells of revolution.

The basic components have been developed in detail in the preceding chapters, with the geometry described in Sections 2.1 and 2.2; the displacements and loading in Section 2.3; the kinematic laws in Section 2.4; the constitutive laws in Section 2.5; and the boundary conditions in Section 2.6.

Finite element analysis of shells of revolution

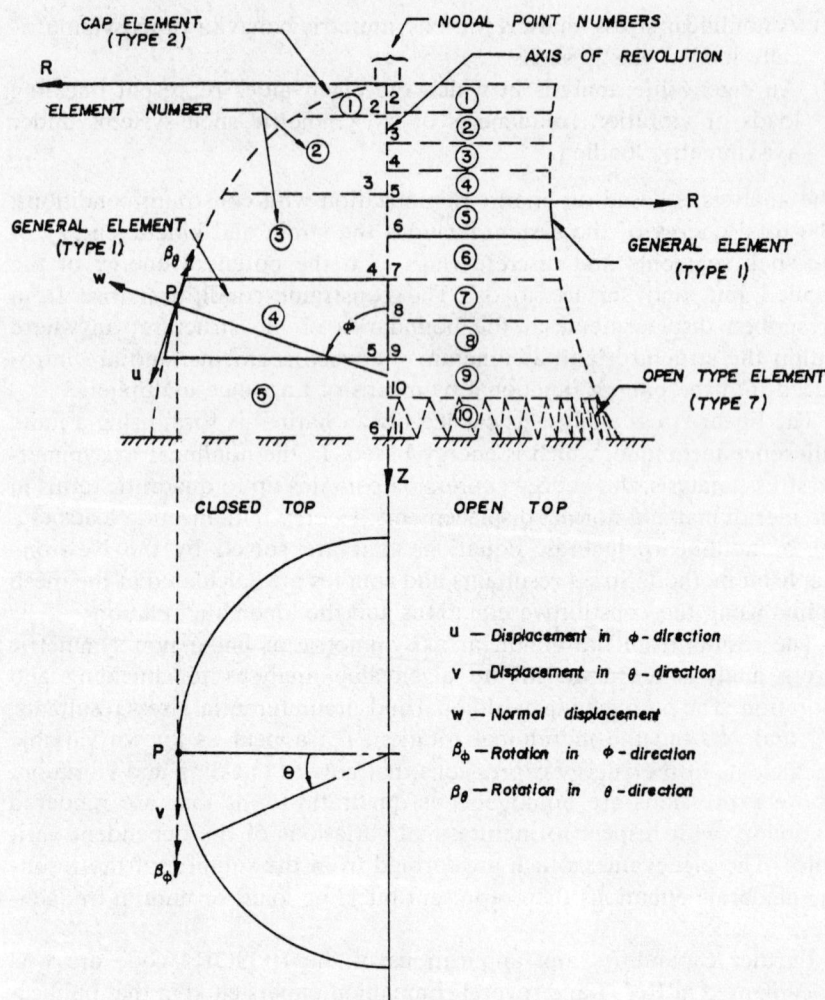

Fig. 7.1 Discretization of rotational shells

Special capabilities of the code for static problems include high order comparison functions which are described in Section 3.1; static condensation in Section 3.2; stiffened shells in Section 3.3.2; compound shells in Section 3.3.3; and open-type elements in Section 3.3.5.

Additional special capabilities for dynamic problems include kinematic condensation described in Section 4.1.2; consistent mass matrices in Sections 4.1.1 and 4.3.2; uniform base motion in Section 4.4.3; response spectrum analysis in Section 4.4.3.2; and direct integration in Section 4.5.

The discretization scheme for both closed and open shells of revolution is shown in Fig. 7.1, and illustrations of many of the capabilities are provided in the following case study.

7.3 Case study of a hyperboloidal shell on column supports

7.3.1 Scope

A hyperboloidal shell of revolution which has dimensions in the range of the tallest modern cooling towers is selected for the case study. The shell profile is shown in Fig. 7.2 where the dimensions and material properties are indicated. The shell is to be analyzed for self-weight, static wind load

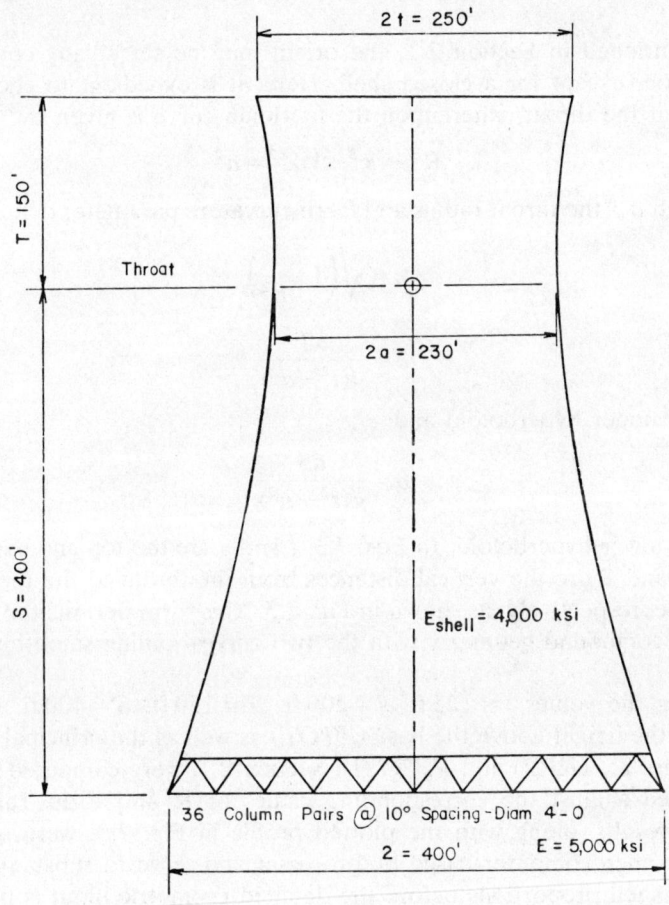

Fig. 7.2 Hyperbolic cooling tower shell

and earthquake loading, which are cases frequently encountered in practice. Stress resultants and stress couples are calculated at strategic locations around the circumference and throughout the height, including the influence of the columns that form the lower boundary.

The steps in the analysis are as follows:

(1) Establishment of meridional geometry.
(2) Selection of discretization pattern.
(3) Determination of loading on shell.
(4) Analysis for individual loading cases with a uniformly supported lower boundary.
(5) Modification for discretely supported lower boundary.

7.3.2 Geometry

As mentioned in Section 2.1, the origin may be set at any convenient elevation, except for a closed shell. Here, it is expedient to choose the origin at the throat, whereupon the meridian curve is given by[3]

$$R^2 - (k^2 - 1)Z^2 = a^2 \qquad (7.1)$$

in which a = the throat radius and k = a curvature parameter defined by

$$k = \sqrt{\left(1 + \frac{a^2}{b^2}\right)} \qquad (7.2)$$

where

$$b = \frac{aT}{\sqrt{(t^2 - a^2)}} \qquad (7.3a)$$

for the upper hyperboloid and

$$b = \frac{aS}{\sqrt{(s^2 - a^2)}} \qquad (7.3b)$$

for the lower hyperboloid. In Eqs. 7.3, t and s are the top and base radii, and T and S are the vertical distances from the throat to the top and to the base, respectively, as shown in Fig. 7.3. This form permits the shell to have a compound geometry with the two curves joining smoothly at the throat.

Using the values $t = 125$ ft, $s = 200$ ft, $T = 150$ ft, $S = 400$ ft and $a = 115$ ft, the height above the base ($ZTOT$) as well as the principal radii of curvature R_ϕ (RFE) and R_θ (RTH), Section 2.1, were computed and are tabulated against the corresponding values of R and Z in Table 7.2. These results, along with the plotted profile in Fig. 7.3, were obtained using a microcomputer-based preprocesser and serve to substantiate the overall shell proportions before the detailed geometric input is prepared for the SHORE program.

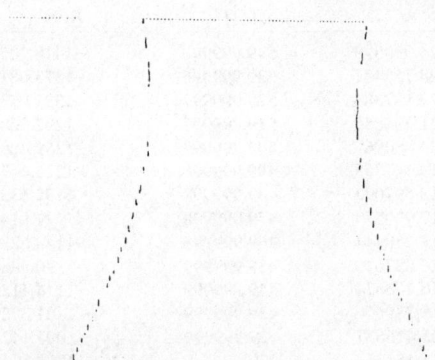

Fig. 7.3 Microcomputer plotted shell profile

Having verified the general configuration, we may now derive the curve of the meridian. The general form of the equation used in the SHORE program is

$$AZ^2 + BRZ + CR^2 + DZ + ER + F = 0 \qquad (7.4)$$

with

$$B = D = E = 0$$

for this case. Therefore we have

$$AZ^2 + CR^2 + F = 0 \qquad (7.5)$$

which may easily be matched with Eq. 7.1. However, it proves to be convenient to retain the equation in the form of Eq. 7.5; to select F as an arbitrary value, say -100; and then to evaluate A and C by using the points $(R, Z) = (125, -150)$ and $(115, 0)$ for the upper curve, and $(200, 400)$ and $(115, 0)$ for the lower curve. The results are

$$0.0008065Z^2 + 0.0075614F^2 - 100 = 0 \quad (Z \le 0) \qquad (7.6a)$$

and

$$-0.0012653Z^2 + 0.0075614R^2 - 100 = 0 \quad (Z \ge 0) \qquad (7.6b)$$

7.3.3 *Discretization pattern*

On Fig. 7.4 the nodal circles delineating the axisymmetric elements along the vertical axis are indicated. Each line indicates a *nodal circle*, and the spacing is chosen so as to fully exploit the efficiency of the high-precision elements, as discussed in Section 3.5.2. The column-supported lower boundary is modeled with open-type (Code 7) elements, described in Section 3.3.5 and shown in Fig. 3.6, with fully continuous (Code 1) members, as denoted in Table 3.1.

Table 7.2 Geometrical data (ft)

Z	R	ZTOT	RFE	RTH
−149.999996	124.999999	549.999996	−1418.70582	126.019839
−139.999996	123.756481	539.999996	−1373.08176	124.654205
−129.999997	122.587383	529.999997	−1331.05339	123.369165
−119.999997	121.494855	519.999997	−1292.52846	122.167262
−109.999997	120.48096	509.999997	−1257.42019	121.050971
−99.9999974	119.547758	499.999998	−1225.64749	120.022682
−89.9999977	118.697093	489.999998	−1197.13527	119.084675
−79.9999979	117.93077	479.999998	−1171.81464	118.239098
−69.9999982	117.250444	469.999998	−1149.62326	117.487947
−59.9999984	116.657619	459.999999	−1130.50545	116.833043
−49.9999987	116.153634	449.999999	−1114.41252	116.276012
−39.999999	115.73965	439.999999	−1101.30291	115.818268
−29.9999992	115.416637	429.999999	−1091.14237	115.460989
−19.9999995	115.185358	420	−1083.90406	115.205112
−9.99999973	115.046367	410	−1079.56879	115.051312
1.3387762E−08	115	400	−1078.125	115
10	115.072735	390	−688.731401	115.084903
20	115.290665	380	−693.307695	115.339235
30	115.652969	370	−700.957256	115.76188
39.9999999	116.158297	360	−711.713545	116.351004
49.9999999	116.804792	350	−725.623059	117.104096
59.9999999	117.590125	340	−742.744982	118.018014
69.9999999	118.511537	330	−763.150857	119.089057
79.9999999	119.565881	320	−786.924037	120.31303
89.9999998	120.749676	310	−814.159221	121.685316
99.9999998	122.059156	300	−844.961861	123.200961
110	123.490321	290	−879.44758	124.854743
120	125.038994	280	−917.741568	126.641252
130	126.700865	270	−959.977939	128.554952
140	128.471543	260	−1006.29918	130.590254
150	130.346593	250	−1056.8555	132.741564
160	132.321578	240	−1111.80422	135.003335
170	134.392092	230	−1171.30929	137.370113
180	136.55379	220	−1235.54071	139.836565
190	138.802411	210	−1304.67403	142.397511
200	141.133801	200	−1378.88988	145.047947
210	143.543928	190	−1458.37354	147.783059
219.999999	146.028892	180.000001	−1543.31454	150.598232
229.999999	148.58494	170.000001	−1633.90621	153.489063
239.999999	151.208465	160.000001	−1730.34539	156.451354
249.999999	153.896018	150.000001	−1832.83222	159.481127
259.999999	156.644302	140.000001	−1941.56962	162.574608
270	159.450177	130.000001	−2056.76314	165.728228
279.999999	162.310658	120.000001	−2178.62088	168.938622
289.999999	165.222908	110.000001	−2307.35295	172.202613
299.999999	168.184236	100.000001	−2443.17145	175.51721
309.999999	171.192097	90.0000008	−2586.29034	178.879601
319.999999	174.244081	80.0000009	−2736.92523	182.287143
329.999999	177.337909	70.0000009	−2895.29308	185.737346
339.999999	180.47143	60.0000008	−3061.61241	189.227882
349.999999	183.642613	50.0000008	−3236.10284	192.756559
359.999999	186.849537	40.000001	−3418.98494	196.321316
369.999999	190.090398	30.000001	−3610.48077	199.920228
379.999999	193.363485	20.000001	−3810.81276	203.551479
389.999999	196.66719	10.000001	−4020.20458	207.213372
399.999999	200	9.53674317E−07	−4238.88076	210.904312

Computer programs and case study

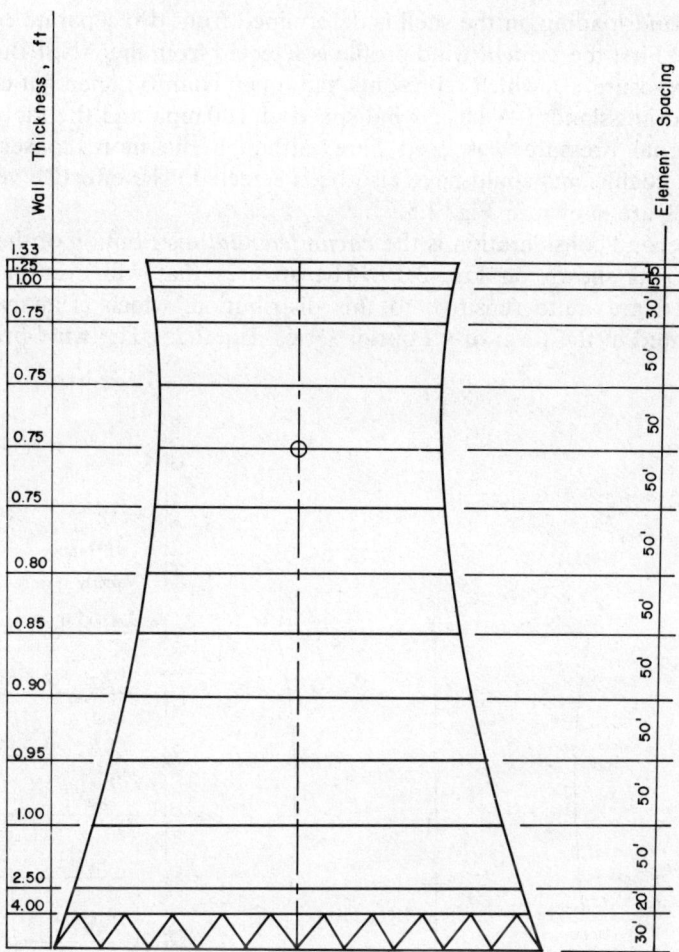

Fig. 7.4 Discretization of hyperboloidal shell

The wall thickness of the shell at each nodal circle is also indicated on Fig. 7.4. The transition in thickness is assumed to be linear for the purposes of analysis, although it might actually occur in 5 to 6 ft steps, which are the heights of construction *lifts*.

7.3.4 Loading

The shell *self-weight* is easily computed from the wall thicknesses shown in Fig. 7.4, taking the unit weight of concrete as 150 lb/ft^3.

Finite element analysis of shells of revolution

The *wind* loading on the shell is determined from two separate considerations. First the *vertical* wind profile is selected from the ANSI Building Code Exposure C,[4] which represents 'flat, open country, open flat coastal belts and grasslands'. A basic wind speed of 100 mph and the tabulation for internal pressure was used here, although the more conservative external coefficients could have also been selected. The effective velocity pressures are shown in Fig. 7.5.

The second consideration is the *circumferential* distribution of the wind pressure, as shown on Fig. 3.19. The stresses that will eventually be computed are quite sensitive to this distribution, which is analytically represented in the form of a Fourier series, Eq. 2.25. The wind pressure

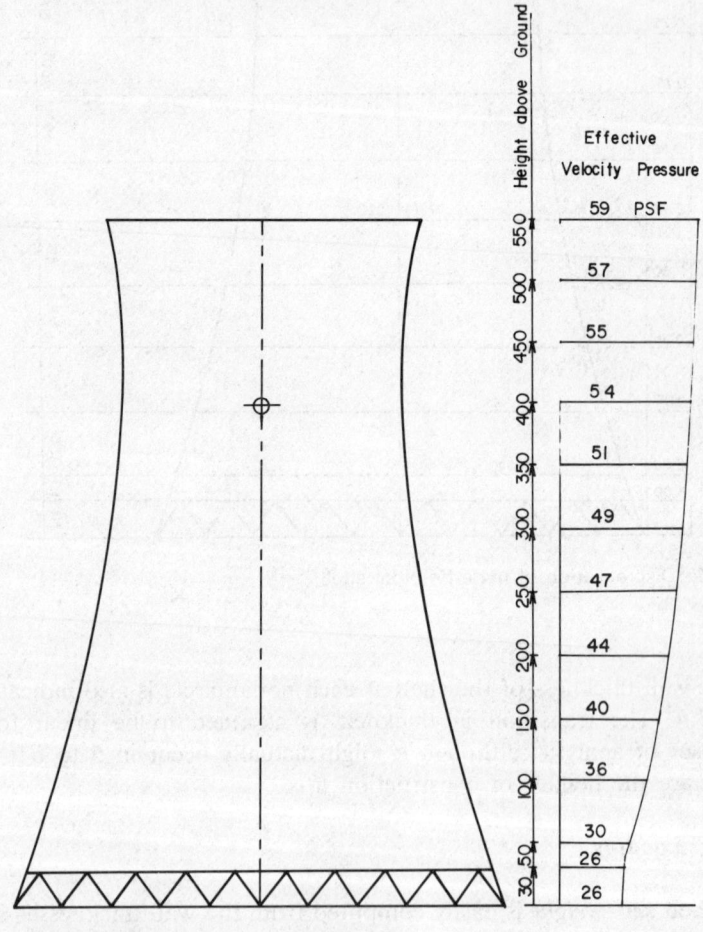

Fig. 7.5 Wind loading

is assumed to act only in the normal direction, which corresponds to f_w in Eqs 2.25 and 2.26, and each harmonic component f_w^j is computed as the product of the basic velocity pressure, shown in Fig. 7.5, and the normalized Fourier coefficient, a^j.

For this example, the Fourier coefficients are taken from full-scale measurements, as suggested in Ref. 5, and are given in Table 7.3 along with the computed harmonic components of the loading, f_w^j. It may be noted that the Fourier coefficients are taken as constant with height in accordance with current practice.

The *earthquake* loading on the shell is calculated by the response spectrum method, as detailed in Section 4.4.3.2. The spectrum selected is shown in Fig. 7.6(a),[6] which is a tripartite plot of the displacement, velocity and acceleration spectra (an example of the latter is shown separately in Fig. 4.1) for 2% of critical damping. The SHORE program is equipped to interpret piecewise linear response spectra, so that the envelope shown on the figure is used with the key coordinates converted to frequency and spectral velocity (ft/s) units.

Following the procedure detailed in Section 4.6.3, the first step is to perform a free vibration analysis. Frequencies for circumferential harmonics $j = 0$ to $j = 10$ are given in Table 7.4(a). These are plotted in Fig. 7.6(b) and show the characteristic minimum in the $j = 3$ to 6 range of harmonics, as observed in Section 4.6.1 and illustrated in Fig. 4.5. In the

Table 7.3 Fourier coefficients and harmonic loading components, f_w^j

			Harmonic					
Node	Height (ft)	Velocity pressure (lb/ft^2)	j a^j	0 -0.3923	1 0.2602	2 0.6024	3 0.5046	4 0.1064	5 -0.0948
14	30 or less	26		−10.1998	6.7652	15.6624	13.1196	2.7664	−2.4648
13	50	30		−11.769	7.806	18.072	15.138	3.192	−2.844
12	100	36		−14.1228	9.3672	21.6864	18.1656	3.8304	−3.4128
11	150	40		−15.692	10.408	24.096	20.184	4.256	−3.792
10	200	44		−17.2612	11.4488	26.5056	22.2024	4.6816	−4.1712
9	250	47		−18.4381	12.2294	28.3128	23.7162	5.0008	−4.4556
8	300	49		−19.2227	12.7498	29.5176	24.7254	5.2136	−4.6452
7	350	51		−20.0073	13.2702	30.7224	25.7346	5.4264	−4.8348
6	400	54		−21.1842	14.0508	32.5296	27.2484	5.7456	−5.1192
5	450	55		−21.5765	14.311	33.132	27.753	5.852	−5.214
4	500	57		−22.3611	14.8314	34.3368	28.7622	6.0648	−5.4036
3	530	58.2		−22.8319	15.1436	35.0597	29.3677	6.1925	−5.5174
2	545	58.8		−23.0672	15.2998	35.4211	29.6705	6.2563	−5.5742
1	550	59		−23.1457	15.3518	35.5416	29.7714	6.2776	−5.5932

Finite element analysis of shells of revolution

case of earthquake loading with uniform horizontal base motion, only the $j = 1$ harmonic participates. The corresponding periods, T_r^1, and the spectral ordinates, scaled from Fig. 7.6(a) and converted to ft/s units, are given in Table 7.4(b).

Also obtained from the free vibration analysis are the mode shapes, which are normalized for each meridional mode r by the normal (w)

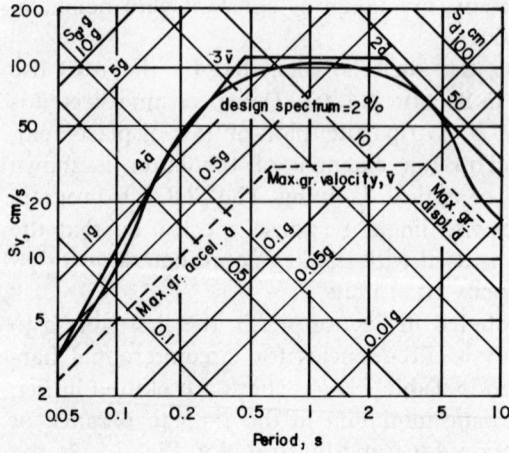

LINEARIZED COORDINATES	
$\omega/2\pi$ (Hz)	Sv (ft/sec)
20	0.12
5.88	1.32
2.22	3.96
0.36	3.96
0.10	1.12

Fig. 7.6(a) Response-spectrum (Ref. [6])

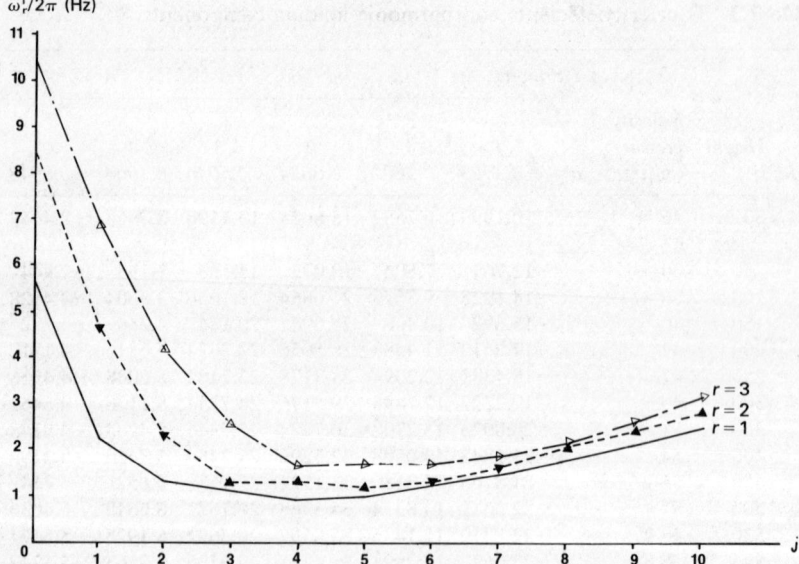

Fig. 7.6(b) Natural frequencies

Table 7.4(a)

j	Natural frequencies (Hz)		
	$\omega_1^i/2\pi$	$\omega_2^i/2\pi$	$\omega_3^i/2\pi$
0	5.754	8.652	10.484
1	2.202	4.672	6.887
2	1.281	2.326	4.130
3	1.092	1.261	2.550
4	0.842	1.311	1.654
5	0.956	1.157	1.654
6	1.230	1.261	1.674
7	1.477	1.634	1.780
8	1.762	2.066	2.106
9	2.090	2.460	2.602
10	2.463	2.887	3.160

Table 7.4(b)

Meridional mode r	$\omega_r^1/2\pi(Hz)$	$T_r^1(s)$	$S_{Vr}^1(ft/s)$
1	2.202	0.454	3.96
2	4.672	0.214	2.19
3	6.887	0.145	1.23

displacement component, Figs 7.7–7.9. The characteristics are similar to those shown for $j = 1$ in Fig. 4.11, with the mode 2 normal displacement in the lower regions of the shell reflecting the flexible, column-supported base condition.

Once the free vibration analysis is completed and the spectral ordinates are determined, the forces acting on the shell are found from Eq. 4.43. This is carried out in the course of the computer solution and the values are not tabulated here.

7.3.5 Self weight stress analysis

The stress resultants for the self dead load are determined following the static analysis procedure developed in Sections 3.1 and 3.2. In this model, a continuous boundary representation of the system of column supports is provided with the open-type element as described in Section 3.3.5.

The meridional and circumferential stress resultants, N_ϕ and N_θ, are

Finite element analysis of shells of revolution

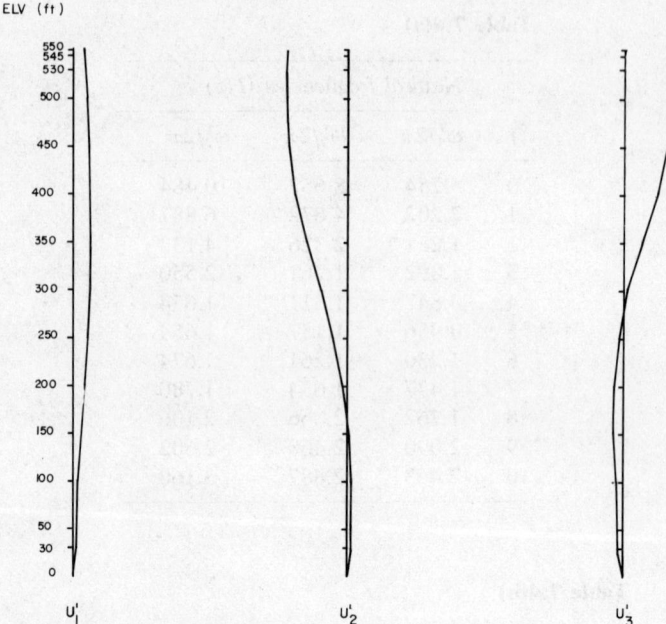

Fig. 7.7 Meridional mode shapes

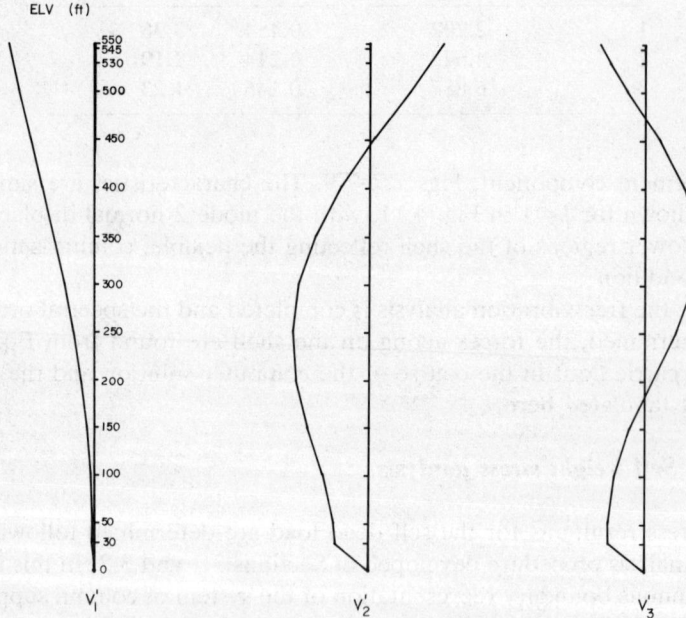

Fig. 7.8 Circumferential mode shapes

Computer programs and case study

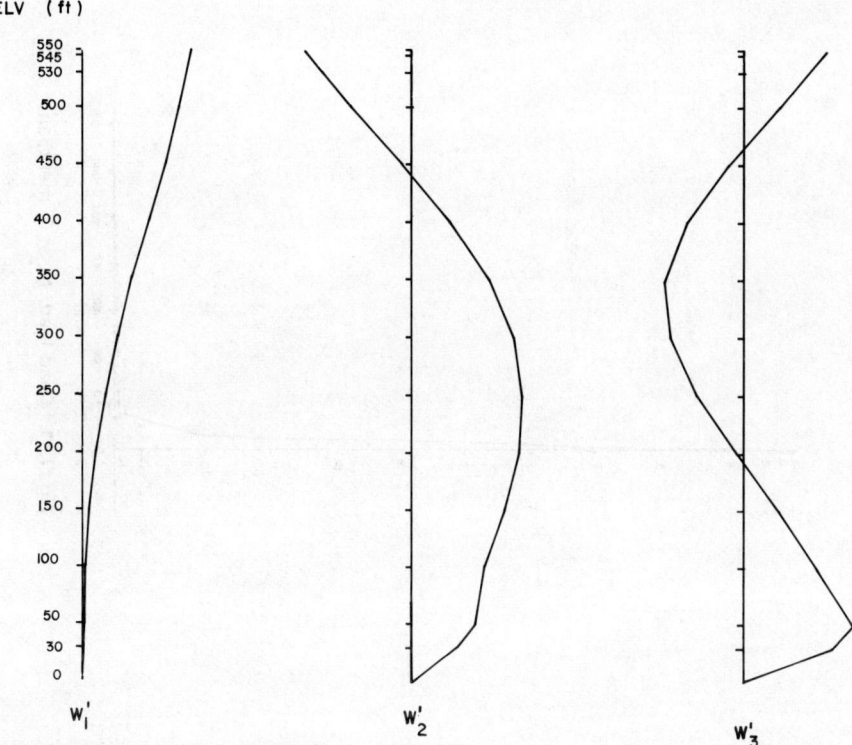

Fig. 7.9 Normal mode shapes

plotted as solid lines in Figs 7.10 and 7.11 and the corresponding stress couples, M_ϕ and M_θ, are shown in Figs 7.12 and 7.13 as a function of shell height. Also, in Table 7.5, the column forces and moments identified on Fig. 3.8 are given.

While the open-type element provides an overall approximation to the flexible boundary, it does not reflect the local amplification of the stress resultants in the vicinity of the column-shell junction. An approximation to this effect is provided by the static correction described in Section 3.3.6. The data for this analysis is tabulated in Table 7.6. Recall that for a symmetrical case, such as self weight, only harmonics which are an even multiple of the number of support points participate. The corrected stress resultants and couples are shown on Figs 7.10–7.13 as dashed lines, confirming the local but severe nature of the amplification. Eight harmonics were processed for this example, but several more would be necessary to provide close convergence to the square waveform shown in Fig. 3.11(a). This is largely a cosmetic problem and, in practice, it has

Finite element analysis of shells of revolution

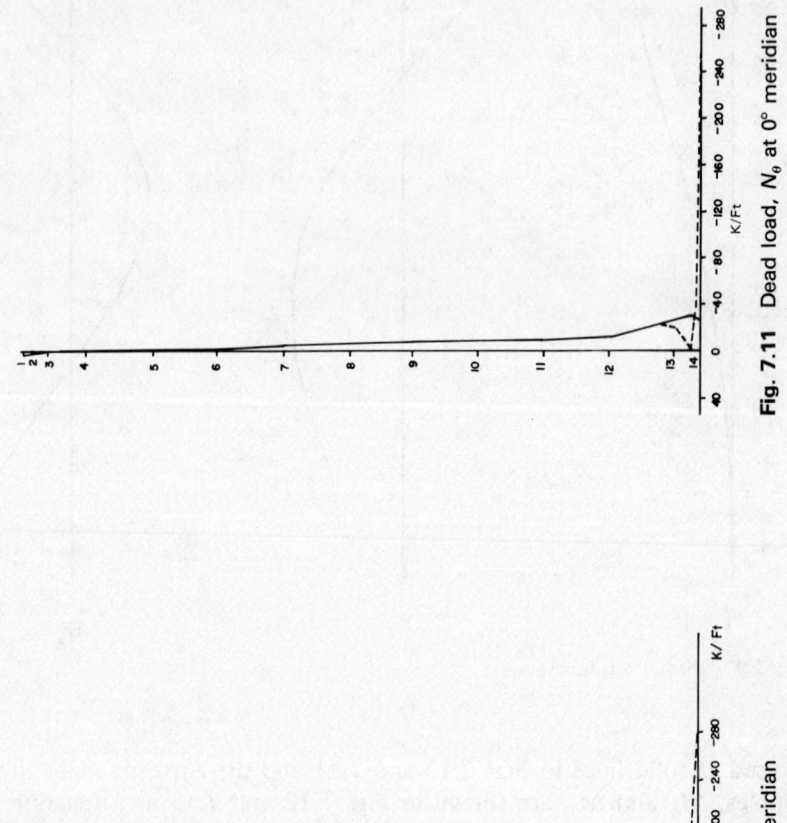

Fig. 7.11 Dead load, N_θ at 0° meridian

Fig. 7.10 Dead load, N_ϕ at 0° meridian

Table 7.5 Column forces (kips) and moments (ft-kips) due to dead load

F_x	F_y	F_z	M_x	M_y	M_z
−1245.71	0	∓5.14	∓7.44	+76.66 (top) +107.88 (bottom)	∓115.77

proved expedient to use a less severe reaction distribution, such as a parabolic curve. This tends to reduce the total number of significant harmonics participating in the solution.

7.3.6 Wind load stress analysis

The stress resultants for the static wind loading are determined for each harmonic from the components given in Table 7.3, and are then evaluated at strategic locations around the circumference using Eq. 2.23.

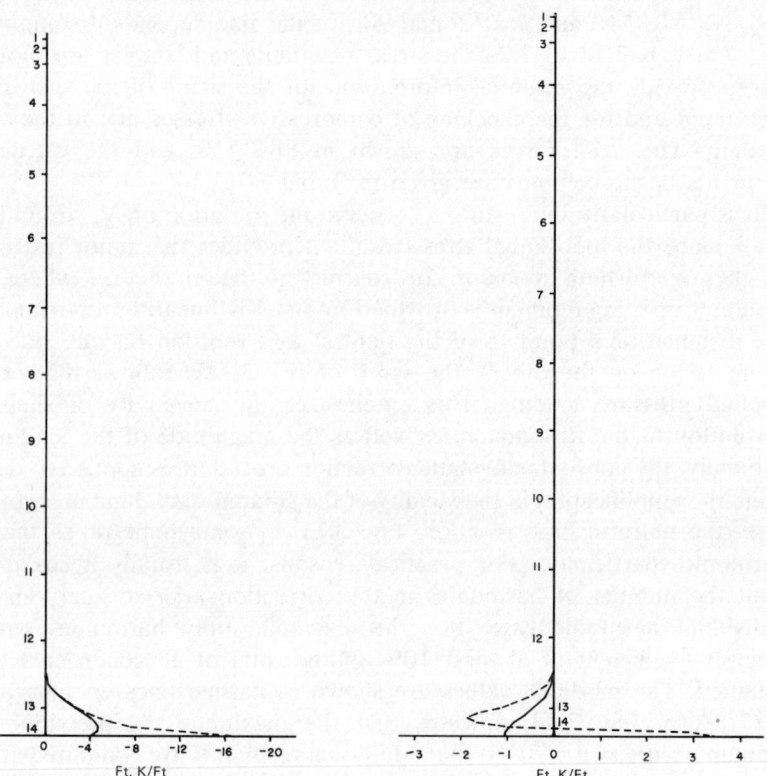

Fig. 7.12 Dead load, M_ϕ at 0° meridian **Fig. 7.13** Dead load, M_θ at 0° meridian

Table 7.6 Data for static correction

i	N_i	Continuous boundary resultants, N_i (Eq. 3.79)	Fourier coefficients, $\tilde{N}_i^j$ (Eq. 3.81) $j=36$	72	108	180	216	252	360	396
1	N_ϕ	−65.77	−119.32	−86.69	−44.20	20.91	28.54	20.61	−16.69	−14.01
2	$N_{\theta\phi}$	0								
3	Q_ϕ	0.275	0.500	0.363	0.185	−0.090	−0.120	−0.086	0.070	0.059
4	M_ϕ	−4.006	−7.267	−5.280	−2.692	1.274	1.738	1.255	−1.017	−0.853
5	M_θ	0								

Number of column pairs = 36. Half column width $\beta = 1.20564°$

The circumferential locations chosen are $\theta = 0°$ (maximum pressure) and $\theta = 72°$ (maximum suction) for the cosine dependent components (N_ϕ, N_θ, M_ϕ, M_θ) and $\theta = 72°$ and $90°$ for the sine dependent component $N_{\theta\phi}$. On Figs 7.14 to 7.23 the stress resultants and couples are plotted. These provide the essential information for the sizing of the shell reinforcement and for the checking of compressive stresses due to the wind loading. The axial forces are shown in Fig. 7.24 and the calculated moments in the columns are given in Table 7.7.

It is particularly interesting to observe the variation of N_ϕ around the shell, since the meridional stress resultant provides the major resistance to the overturning moment. In contrast to beam theory, where the resultant resisting moment is provided by stresses linearly proportional to the distance of a point from the neutral axis (tension on one side and compression on the other), the sense of N_ϕ closely follows that of the applied pressure loading. This emphasizes the sensitivity of shells of revolution to the *distribution*, as well as the magnitude of the loading.

Finally, the approximate static correction procedure is applied to determine the amplification in the vicinity of the column–shell junctures. In this case, the uniform base reaction, Fig. 3.11, is non-symmetric so that all harmonics participate. For practical reasons, it is usually necessary to limit the number of harmonics in the correction *prior* to analyzing the individual harmonic case. For this example, only harmonics with a Fourier coefficient of at least 10% of the total of all coefficients were retained. The modified values are shown as dashed lines on Figs 7.14–7.23. Also, Fig. 7.23 indicates that the maximum compression at a column occurs at $\theta = 70°$ so that additional resultants are computed at this location and are shown in Figs 7.25–7.29. The amplifications in this region are perhaps even larger than at $\theta = 0°$.

7.3.7 Earthquake load stress analysis

The stress resultants for seismic loading are computed from an RSS combination of the individual harmonic results, as discussed in Section 4.4. With the nodal displacements in each mode evaluated in the form of Eq. 4.46, the strains and stress resultants are found mode-by-mode using the procedures of Sections 3.2.3.3 and 3.2.3.4, and then combined by the RSS method.

In Figs 7.30–7.33, the key stress resultants and couples are indicated along with correction for the discrete column reactions, carried out in a similar manner to the wind load case. In addition to the inherent approximations embodied in the correction for discrete column reactions, an additional factor is introduced here, since the loading and primary analysis is dynamically-based while the correction is only static.

In Table 7.8, the axial forces and moments in key columns are given, and, in Fig. 7.34, the axial forces are shown, mode-by-mode and RSS. The higher modes play a significant role in the response, and the possibility for large tensions in the columns (even with some compensating compression provided by the self weight) is evident, as indicated in Table 7.5. Of course, the design response spectrum used for the example, Fig. 7.6(a), is quite severe, but the trend of large forces introduced into the columns will persist, even with less severe spectra.

7.3.8 Influence of upper ring beam

As shown on Fig. 7.4, the shell is gradually thickened towards the upper edge from a minimum thickness of 0.75 ft to 1.33 ft, forming, in effect, an upper ring beam or cornice. The importance of stiffening the free edges of thin cylindrical-type shells to prevent an inextensional ovaling instability under axial compression or normal pressure is well known,[7] but it is interesting to examine the influence of the thickened upper edge on the dynamic characteristics of this shell.

In Table 7.9, some results for the shell with the thickness maintained at a constant value above the throat are given for comparison with Table 7.4(a). With respect to the minimum frequency, corresponding to $j = 4$, the difference is insignificant. For the lower harmonics, the upper ring beam is somewhat *detrimental* in that the constant thickness shell has higher frequencies. This is easy to appreciate for the pure beam mode, $j = 1$, where the frequency is essentially proportional to the stiffness to mass ratio. The upper ring beam simply adds mass in this mode, while not contributing stiffness. For the higher harmonics which are the characteristic shell modes, the ring beam produces only slightly higher frequencies.

Finite element analysis of shells of revolution

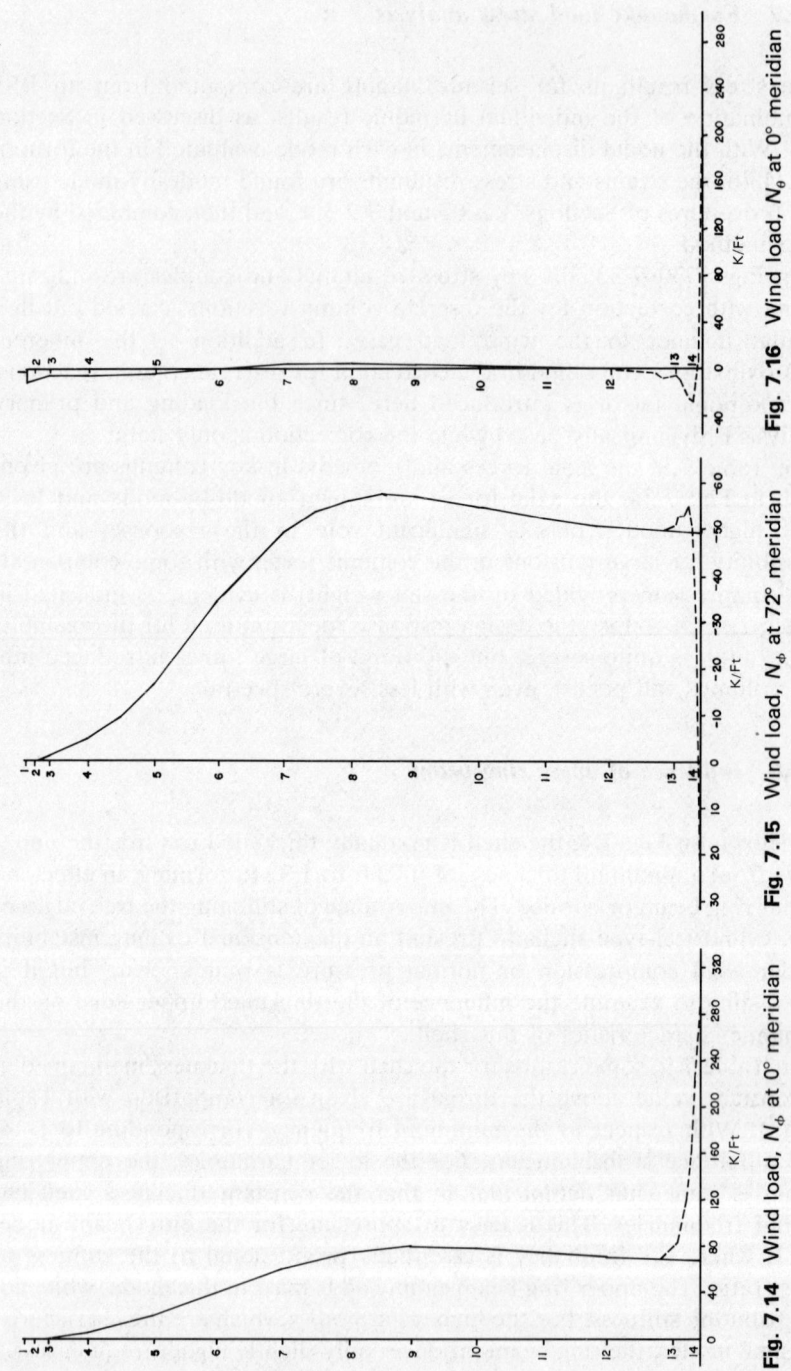

Fig. 7.16 Wind load, N_θ at 0° meridian

Fig. 7.15 Wind load, N_ϕ at 72° meridian

Fig. 7.14 Wind load, N_ϕ at 0° meridian

Computer programs and case study

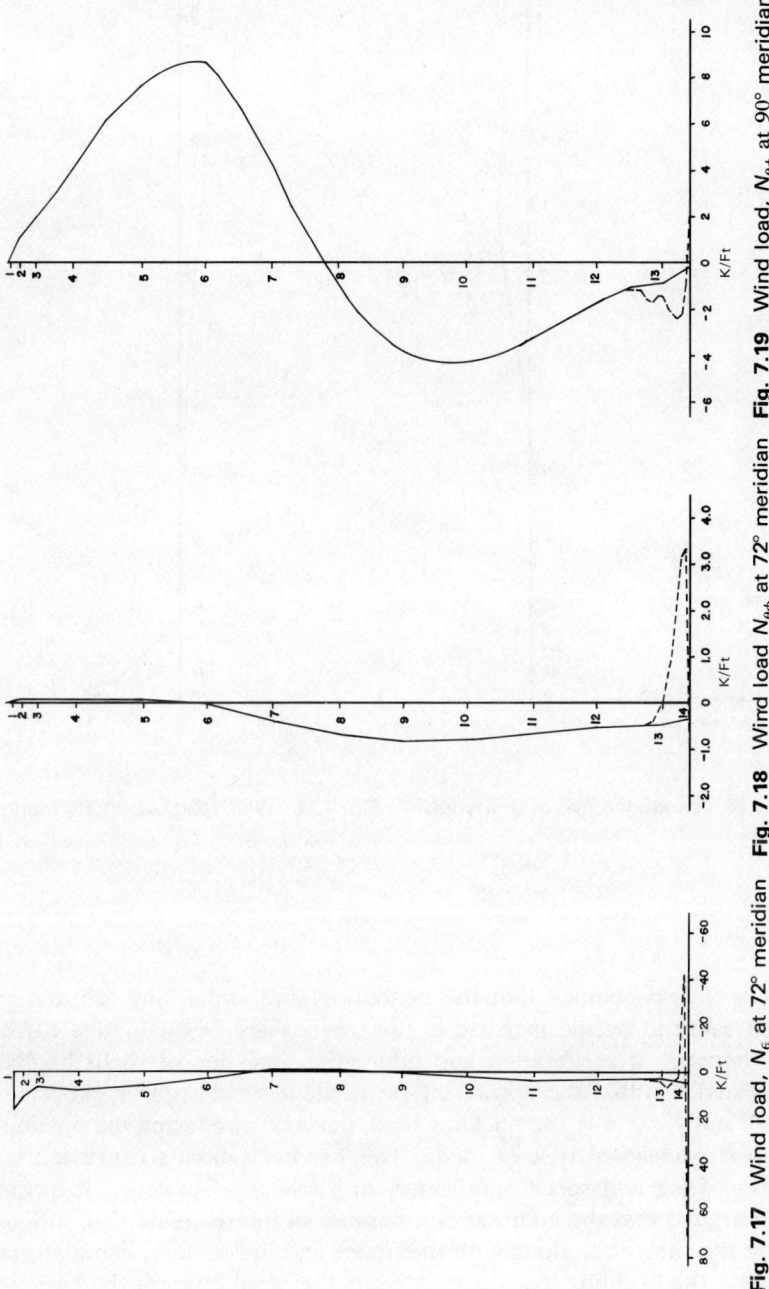

Fig. 7.17 Wind load, N_6 at 72° meridian Fig. 7.18 Wind load $N_{\theta\phi}$ at 72° meridian Fig. 7.19 Wind load, $N_{\theta\phi}$ at 90° meridian

Finite element analysis of shells of revolution

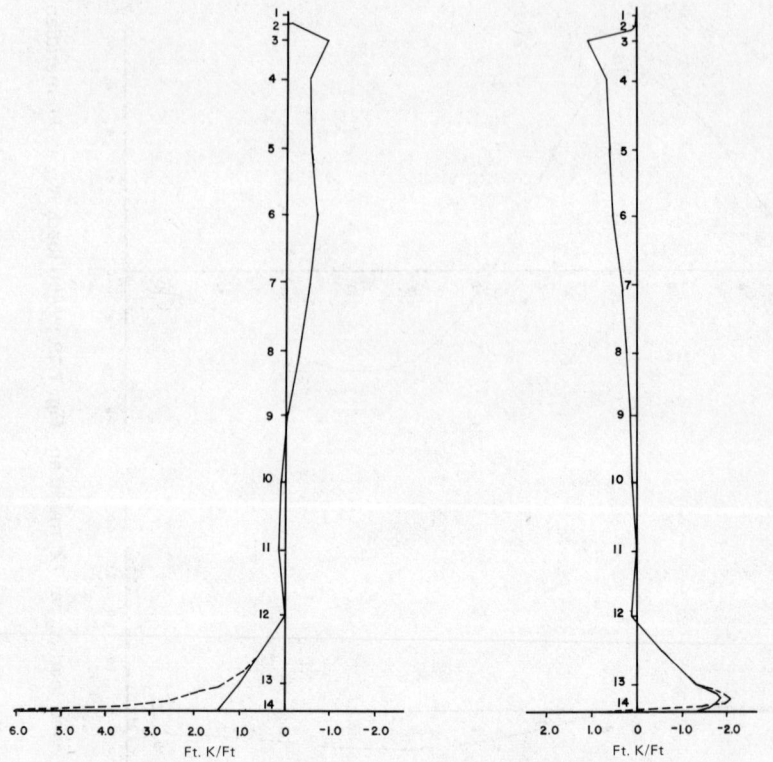

Fig. 7.20 Wind load, M_ϕ at 0° meridian **Fig. 7.21** Wind load, M_ϕ at 72° meridian

It is thus concluded that the benefits of the upper ring are marginal when assessed by the increase in the frequencies. A qualitative correlation between free vibration and bifurcation buckling of shells has been suggested,[8] so that the upper ring beam alone would not be expected to significantly increase the buckling load, outside of reducing the possibility of an inextensional type of mode. This has been discussed previously in Section 5.6.1 and specifically shown in Table 5.1. However, it has also been argued that the addition of a number of intermediate ring stiffeners along the shell axis, along with the upper and lower rings, can materially increase the buckling load; alternatively, this would permit the basic wall thickness, were the shell required to achieve a required buckling capacity without ring stiffeners, to be reduced.[9]

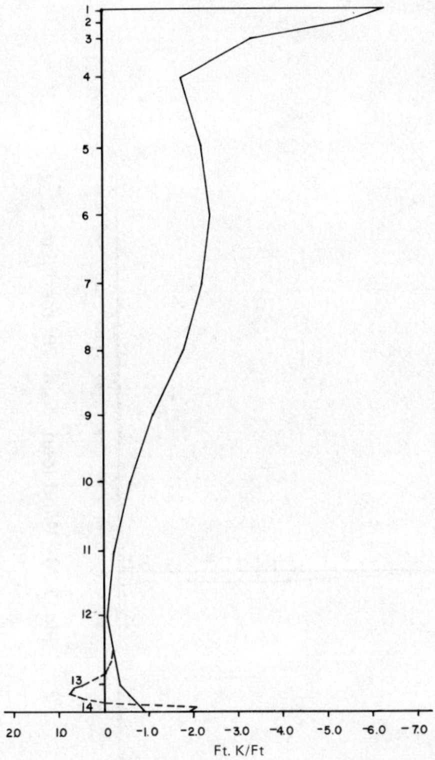

Fig. 7.22 Wind load, M_θ at 0° meridian 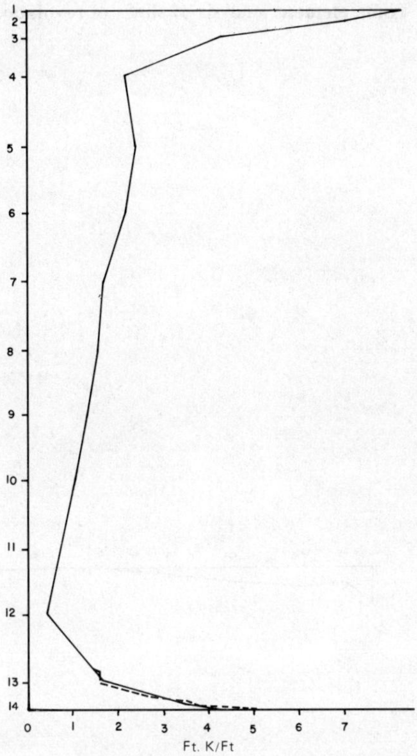   **Fig. 7.23** Wind load, M_θ at 72° meridian

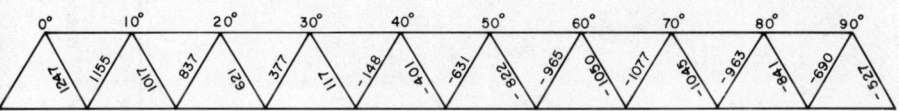

Fig. 7.24 Column axial forces for wind load (kips). −: Comp.; +: Ten

Table 7.7 Column moments due to wind load (ft-kips)

Angle	M_x	M_y (top)	M_y (bottom)	M_z
0°	∓22.34	−16.96	+65.58	±124.61
10°	∓21.80	−7.46	+66.95	±137.30
20°	∓20.09	+12.03	+64.98	±131.37
30°	∓15.91	+34.31	+51.06	±107.56
40°	∓7.49	+48.49	+20.42	±68.94
50°	±4.99	+46.66	−21.15	±22.38
60°	±18.11	+29.05	−58.42	∓21.59
70°	±26.58	+4.04	−75.37	∓51.83
80°	±26.29	−16.99	−65.89	∓61.30
90°	±17.25	−26.73	−38.59	∓50.74

Finite element analysis of shells of revolution

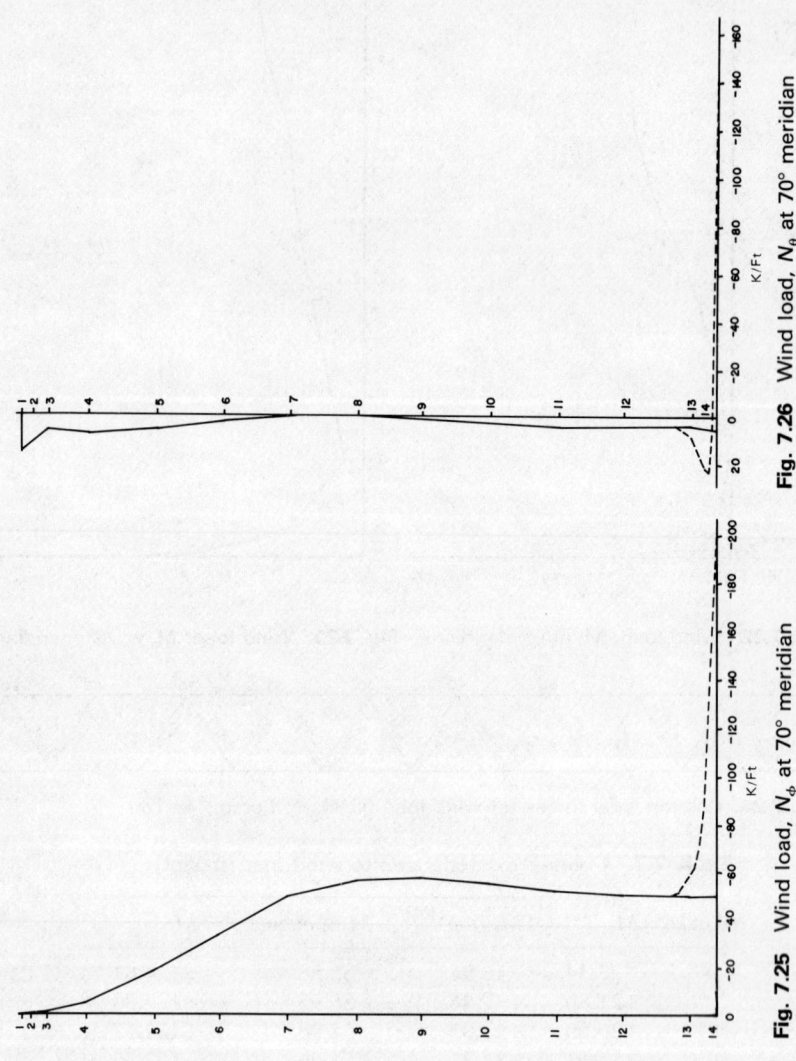

Fig. 7.25 Wind load, N_ϕ at 70° meridian

Fig. 7.26 Wind load, N_θ at 70° meridian

Computer programs and case study

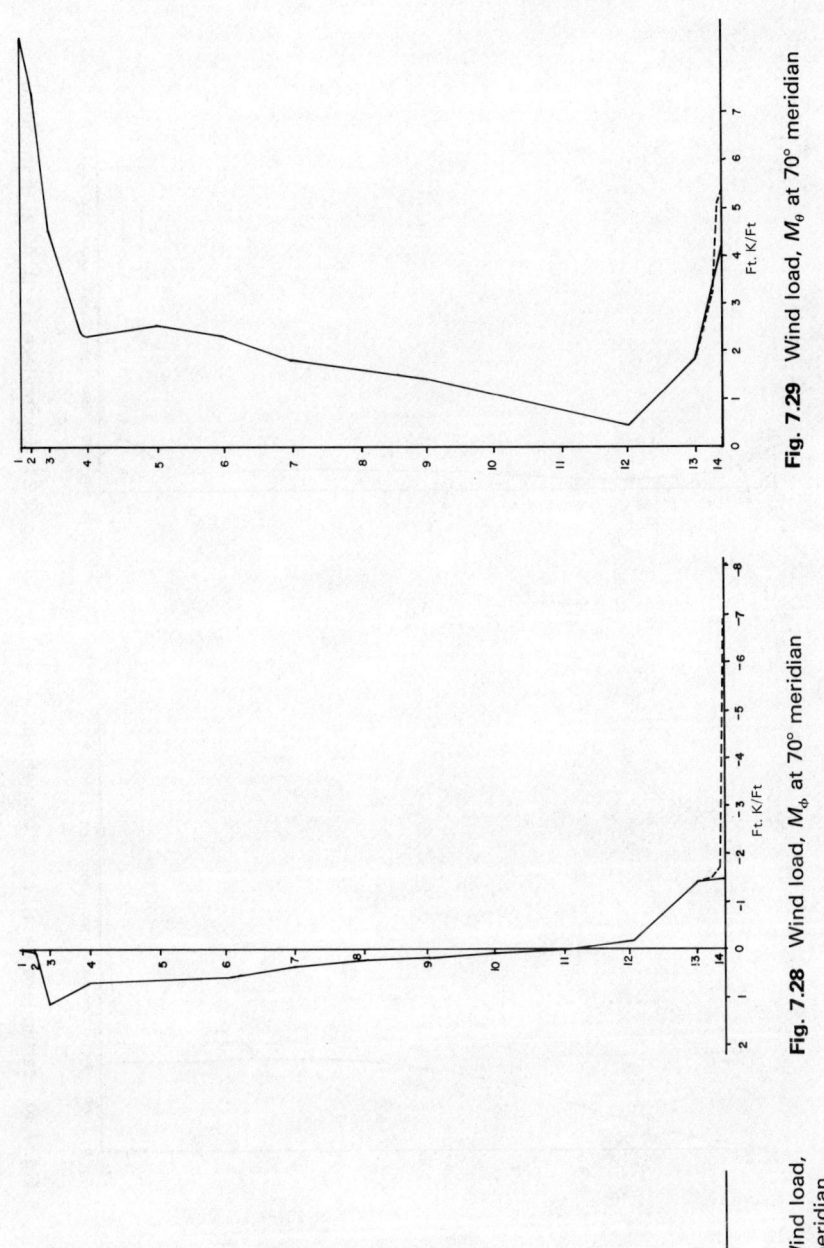

Fig. 7.29 Wind load, M_θ at 70° meridian

Fig. 7.28 Wind load, M_ϕ at 70° meridian

Fig. 7.27 Wind load, $N_{\theta\phi}$ at 70° meridian

Finite element analysis of shells of revolution

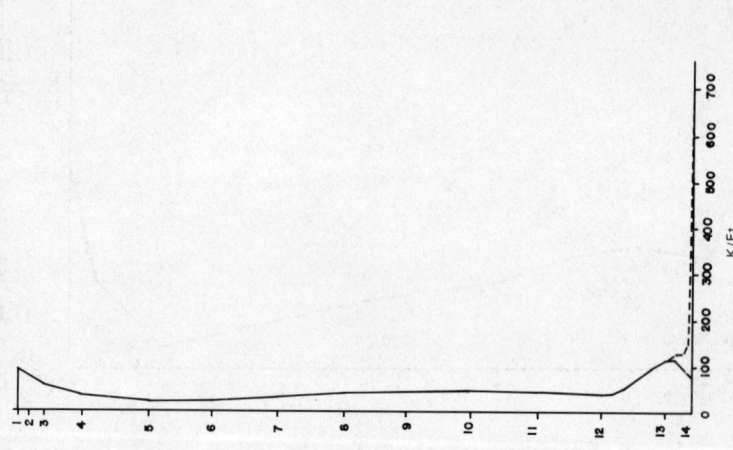

Fig. 7.31 Earthquake ($j=1$), N_θ at 0° meridian

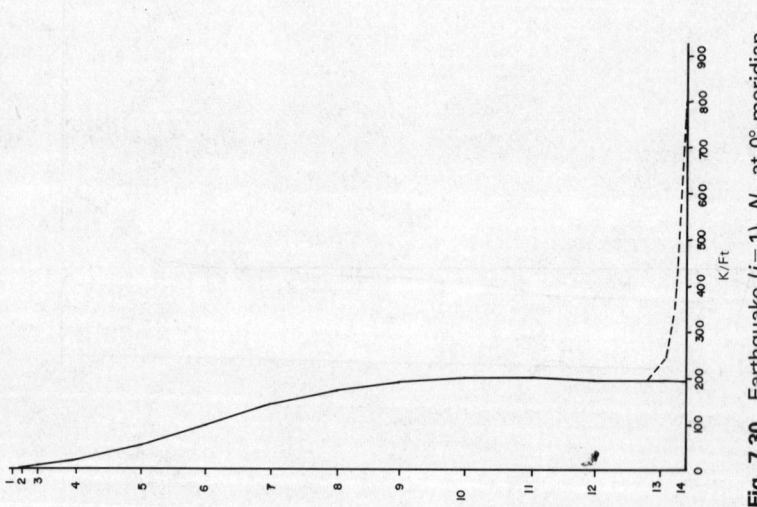

Fig. 7.30 Earthquake ($j=1$), N_ϕ at 0° meridian

Computer programs and case study

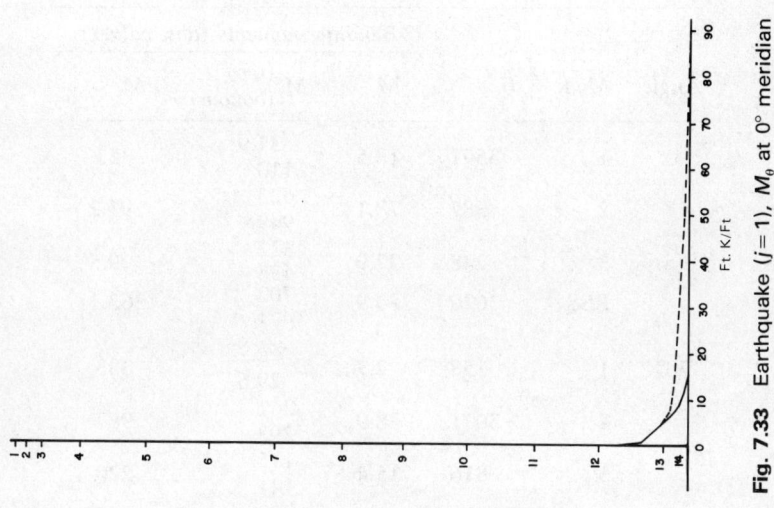

Fig. 7.33 Earthquake ($j=1$), M_θ at 0° meridian

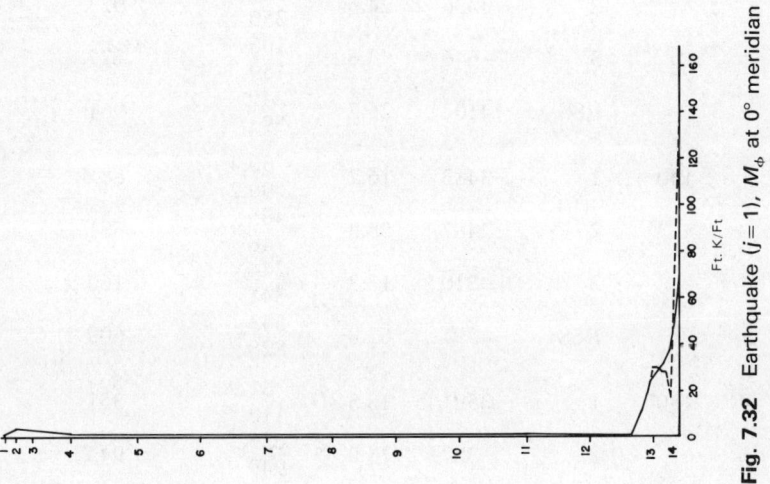

Fig. 7.32 Earthquake ($j=1$), M_ϕ at 0° meridian

Finite element analysis of shells of revolution

Table 7.8 Axial forces, kips and moments (ft-kips) due to earthquake load

				Bending moments (abs. values)		
Angle	Mode	F_x	M_x	M_y (top) (bottom)		M_z
0°	1	3591	18.5	31.9 110		351
	2	387	78.1	623 949		91.2
	3	−248	27.9	322 438		9.5
	RSS	3620	84.9	702 1051		363.1
70°	1	158	2.7	2.2 29.6		383
	2	−3071	28.9	201 293		987
	3	−816	15.4	115 141		220
	RSS	3182	32.9	232 326		1082
110°	1	−2298	9.9	19.6 46.0		143
	2	−3336	24.5	225 356		923
	3	−646	3.6	105 159		227
	RSS	4102	26.7	249 393		961
140°	1	−3483	16.7	30.4 90.2		88.5
	2	−2487	58.3	485 748		581
	3	−310	17.3	244 341		160
	RSS	−4291	63.0	543 828		609
180°	1	−3591	18.5	31.9 110		351
	2	−387	78.0	623 949		94.2
	3	248	27.9	322 438		9.5
	RSS	3620	84.9	702 1051		363

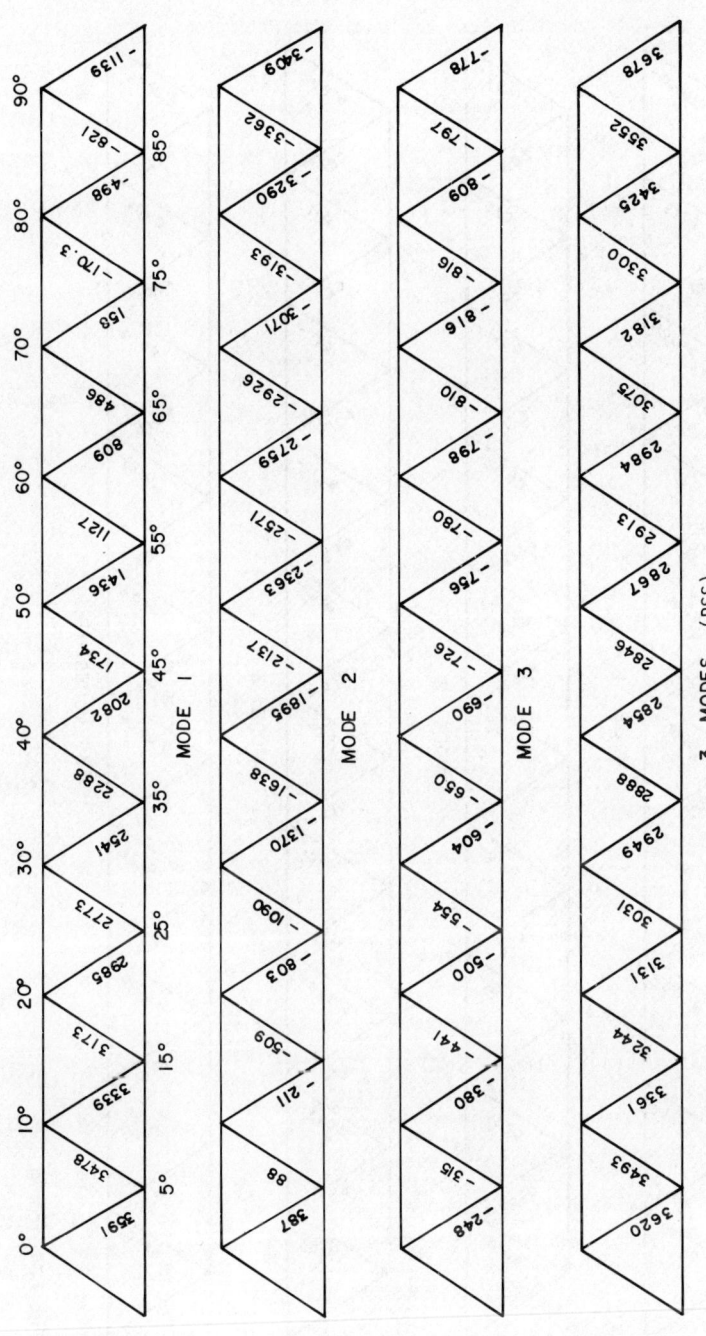

Fig. 7.34 Column axial forces from horizontal response spectrum

Finite element analysis of shells of revolution

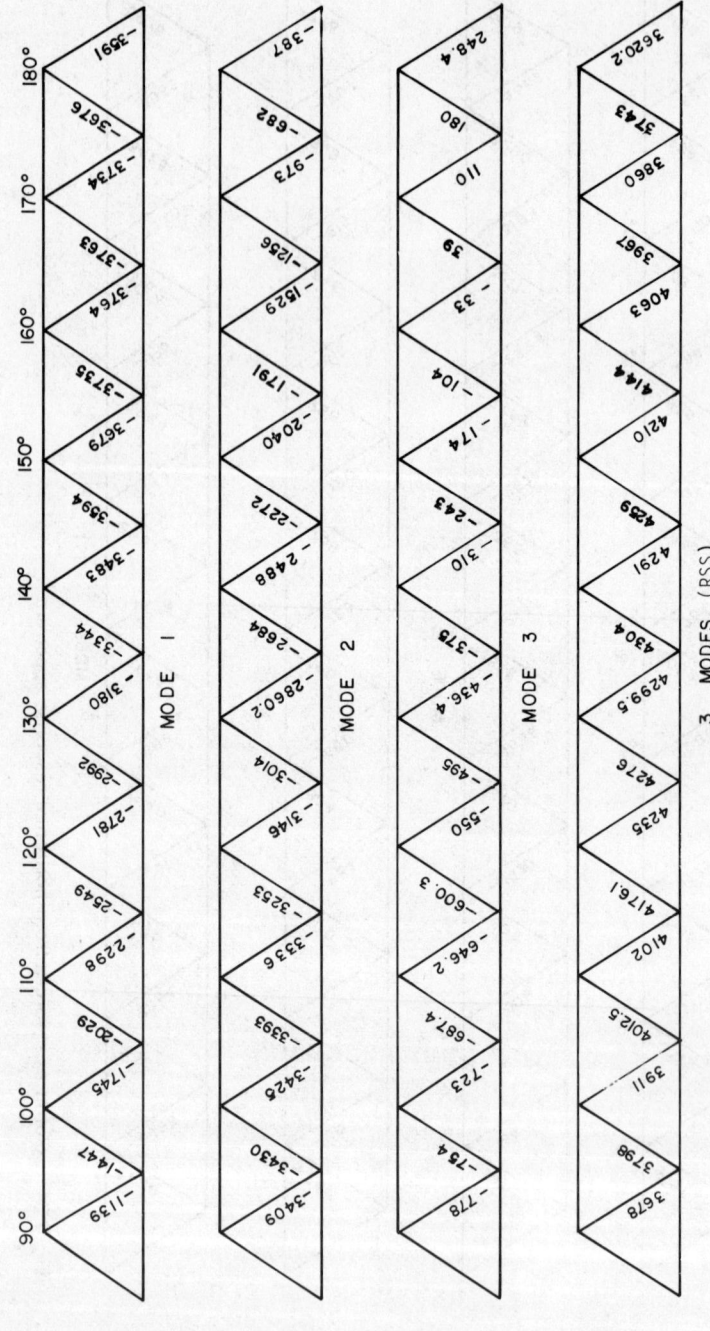

Fig. 7.34 (contd.)

Table 7.9 Free vibration of cooling tower ($h = 0.75$ at top)

j	Natural frequencies (Hz)		
	$\omega_1^i/2\pi$	$\omega_2^i/2\pi$	$\omega_3^i/2\pi$
0	5.939	8.653	10.487
1	2.320	4.755	6.978
2	1.338	2.405	4.193
3	1.156	1.268	2.609
4	0.845	1.310	1.691
5	0.919	1.135	1.649
6	1.152	1.218	1.632
7	1.428	1.572	1.648
8	1.759	1.880	2.049
9	2.090	2.315	2.472
10	2.463	2.803	2.888

References

1. Bushnell, D. 'Stress, Stability and Vibration at Complex Branched Shells of Revolution', *J. Computers and Structures*, Vol. 4, 1974, pp. 399–435.
2. Basu, P. K. and Gould, P. L., 'SHORE-III', Shell of Revolution Finite Element Program, Users' Manual, Structural Division Research Report No. 49, Department of Civil Engineering, Washington University, St. Louis, Mo.
3. Gould, P. L., *Static Analysis of Shells*, Lexington Press, 1977, p. 85.
4. American National Standard Building Code Requirements for Minimum Design Loads in Buildings and Other Structures, American National Standards Institute, ANSI 58.1, New York, NY, 1972, p. 22.
5. 'Reinforced Concrete Cooling Tower Shells—Practice and Commentary', ACI Title 74-2, *J. ACI*, Vol. 74, No. 1, June 1977, pp. 22–31.
6. Clough, R. W. and Penzien, J., *Dynamics of Structures*, McGraw-Hill Book Co., New York, 1975, p. 542.
7. Timoshenko, S. P. and Gere, J. M., *Theory of Elastic Stability*, 2nd edition, McGraw-Hill Book Co., New York, 1961, pp. 461–462.
8. Nguyen, B., Rosemeier, G., Spierig, S. and Stein, E., 'Dynamisches Verhalten von Kühlturmschalen infulge Windkräften unter besonderer Berücksichtigung kinetischer Instabilitäten', *Konstruktiver Ingenieurbun Berichte*, Ruhr-Universität Bochum, W. Germany, Vol. 35–36, pp. 98–103.
9. Krätzig, W. B., Peters, H. L. and Zerna, W., 'Naturzugkühlturme aus Stahlbeton-Derzeitiger Stand und Entwicklungsmöglichkeiten', *Beton-und-Stahlbetonbau*, 1978, Vol. 2, pp. 37–42; Vol. 3, pp. 66–72.

Index

Aas-Jakobsen, A., 52, 75
Abel, J. F., xiii, 123, 126, 129, 132, 134, 139, 143–5
Absolute value sum (ABS), 88
Abu-Sitta, S. H., 85, 96, 97, 128
Acceleration
 base, 89
 ground, 85, 97, 96
 spectral, 87
Ahmad, S., 2, 4, 148, 149, 168
Amplitude, 89
Angle(s), 47
 circumferential. *See* Circumferential angle
 of departure, 48
 of inclination, 47
ANSI building code, 178, 199
Antisymmetrical loading. *See* Loading, antisymmetrical
Arc length, 9, 30
Archer, J. S., 4, 5
Area
 cross-sectional, 80
 surface, 31, 132
Asymmetrical loading on shells. *See* Loading, nonsymmetrical
Axial rigidity, 18, 45
Axial tension, 164–6
Axis of rotation, 1
Axisymmetric shells. *See* Shells of revolution

Banded equations, 33, 41, 152
Bandwidth, 148, 152–7
Bar. *See* Element, bar
Base
 conditions, 96, 98, 114
 fixed, 97 103, 116, 118, 122, 130
 flexible, 96–103, 116
 motion, 85–8, 172
Basu, P. K., xiii, 16, 17, 25, 27, 33, 42, 51, 75, 80, 82, 90, 96, 123, 128, 129, 132–8, 145, 171, 199
Beam, ring, 97, 98
Bending
 moment, 3. *See also* Stress couple
 stiffness, 18
 stress. *See* Stress, bending
 theory, 138
 see also Flexural
Bifurcation buckling, 138–43. *See also* Instability
Billington, D. D., 123, 126, 129, 132, 139, 145

Blast load. *See* Load, blast
BOSOR, 3, 71–3, 169–71
Boundary, 30
 clamped. *See* fixed
 continuous, 53, 101, 109, 174
 discontinuous, 1, 52, 174
 dynamic, 101
 energy transmitting, 102, 105
 fixed, 20
 flexible, 139, 140
 force, 30
 free, 20, 187
 frequency dependent, 101–2
 hinged, 20, 139
 idealizations of, 141
 interelement, 27
 lower, 105, 110
 requirements, 19–20
 roller, 20
 simply-supported. *See* hinged
 sliding, 20
 see also Boundary conditions
Boundary conditions, 1, 19–20, 24, 33, 78, 139, 171
 displacement, 22, 33
 idealized, 1
 interelement, 24
 lower, 102
 mixed, 19
 natural, 19
 physical, 24
 shells of revolution, for, 19–20
 stress, 22
 see also Boundary
Branched shells, 2, 36, 40–1, 80, 153, 169
Brombolich, L., xiii, 4, 6, 9, 24
Bubble functions, 27
Buchert, K., 37, 75
Buckling, 66–8, 138–43, 190. *See also* Instability
Bulge imperfection. *See* Imperfection, bulge
Bushnell, D., xiii, 2–5, 14, 17, 25, 36, 41, 68, 75, 76, 138, 143, 145, 169, 171, 199

Calculus of Variations, 24
Carter, R. L., 91, 92, 128
Chan, A. S. L., 4, 5
Chang, S-C., 143–5
Change(s) in curvature, 11, 13, 16, 17, 22
 displacement equations, 17
Characteristic value. *See* Eigenvalue

201

Index

Cheung, Y. K., 3, 5
Circular cylindrical shells. *See* Cylindrical shells
Circular ring. *See* Beam, ring
Circumferential angle, 50
Circumferential bending moment. *See* Stress couple, circumferential
Circumferential boundary, 21
Circumferential change in curvature. *See* Change(s) in curvature
Circumferential coordinate, 7, 31
Circumferential deformation. *See* Displacement, circumferential
Circumferential displacement. *See* Displacement, circumferential
Circumferential force. *See* Stress resultant
Circumferential length, 12
Circumferential loading. *See* Loading, circumferential
Circumferential moment. *See* Stress couple, circumferential
Circumferential radius of curvature. *See* Principal radius of curvature
Circumferential ring, 21
Circumferential rotation. *See* Rotation, circumferential
Circumferential strain. *See* Strain, circumferential
Circumferential stiffening. *See* Stiffening, circumferential
Circumferential transverse shearing strain. *See* Transverse shearing strain, circumferential
Circumferential variable, 4, 12
Clough, R. W., 35, 75. 84, 89, 128, 179, 199
Codazzi. *See* Gauss–Codazzi relations
Coffin, G. K., 3, 5, 66, 76
Cole, P. B., 132, 139, 145
Collar. *See* Stiffener, ring
Column(s)
 flexibility of, 97, 100, 116
 forces in, 108, 109, 111, 122, 185–7, 191, 196–8
 inclined, 41, 42, 95–100
 mass of, 96
 moments. *See* forces in
 reactions, 52–4, 102
 representation of. *See* Elements, open-type
 tension in 187
Column-supported shell(s), 13, 19, 41, 52, 95–100, 173–99
Combined shells. *See* Compound shells
Comparison function(s), 4, 24, 26–8, 172. *See also* Polynomials, shape function
Compatibility, 148
Complementary strain energy. *See* Energy, complementary strain
Complex response method, 88–90, 114–15
Compliance, 107
Compound shells, 36, 38–40, 66, 172
Computer
 analysis, 2
 programs, 2, 3, 169–73
Concentrated force(s), 13
Concentrated loading. *See* Loading, concentrated
Concentrated moment(s), 13
Condensation, 81, 82, 154
 kinematic, 78, 83, 86, 90, 92, 172

Condensation—*contd.*
 static, 27, 31–2, 40, 49, 78, 155, 172
Conical frusta, 3
Conical shell(s), 57
Connectivity matrix, 43
Consistent load vectors, 31, 51
Consistent mass matrix. *See* Mass matrix, consistent
Consistent nodal forces, 4
Constant coefficients, 26
Constraint(s), 171
Constitutive law(s), 17–19, 21, 22, 27, 99, 171
 matrix, 44
Continuity
 C^0, 27, 131, 135
 C^1, 27, 135
 displacement, 42
 interelement, 12, 56
 requirements, 2, 26
 slope, 38, 42
Continuum, 1, 2
Convergence, 28, 56, 91, 92, 95
Cooling tower. *See* Hyperbolic cooling tower; Hyperboloidal shell of revolution
Coordinate(s)
 Cartesian, 7, 86, 149
 curvilinear, 35, 38, 46, 85, 86, 112, 114, 149
 cylindrical, 149
 generalized, 84–5, 88, 89, 97, 152
 global, 38, 40
 local Cartesian, 42, 46
 meridional, 8
 orthogonal curvilinear. *See* curvilinear
 polar
 relative, 85
 transformation, 36, 46–8, 81, 84, 152
 see also specific geometrical forms
Core region, 101
Cornice, 139–41
CPU time, 165
Creep, 2
Critical load. *See* Buckling; Instability
Curve
 element. *See* Element, curved
 high-order, 10
 least squares, 10
 low-order, 10
 planar, 7
 plotted, 10
 tabulated, 10
Curvature, 3. *See also* Change in curvature; Gaussian curvature
Cut-outs, 147, 164–7
Cylinder, 38. *See also* Cylindrical shell
Cylindrical shell(s)
 analysis of, 152–7
 axially loaded, 139
 axisymmetrically loaded, 57
 blast load on, 122–5
 buckling of, 139
 edge loaded, 58–61
 fluid-filled, 123–7
 free vibration of, 83, 90–3
 hydrostatic loading on, 59, 62, 63

Index

Cylindrical shells—*contd.*
  thermal load on, 66, 68
  vibration of. *See* free vibration of
  with circular cutout, 164–7
  with torospherical head, 66–74

Damping, 82, 118
  coefficient, 79, 115
  critical, 96
  hysteretic, 115
  of foundation, 116
  proportional, 79
  radiation, 116
  ratio, 79, 85, 87
  viscous. *See* Damping matrix, viscous
  see also Damping matrix
Damping matrix, 79, 89
  element, 78
  generalized, 85, 87
  reduced, 86
  viscous, 22, 77, 78, 86
  see also Damping
Dead-weight loading. *See* Loading, self-weight
Decay length, 56
Deflection. *See* Deformation; Displacement
Deformation(s)
  prebuckling. *See* Prebuckling deformation(s)
  small, 16
Degrees of freedom (DOF), 83, 84, 102, 105, 149, 152–7
Density, 64, 91, 93
Depth ratio, 102, 106
Deviations. *See* Imperfections
Direct integration solution, 79, 90, 115, 172
  stiffness method, 33
  see also Time history
Discontinuity, 3
  slope, 38, 39, 46, 47, 48, 74, 81
Discrete supporting systems. *See* Column-supported shells
Discretization, 95
  criteria, 56–8
  pattern, 58, 70, 172, 174–7. *See also* Mesh
Displacement(s), 14, 22, 27, 56, 86, 97, 150, 171
  amplitude, 115, 118
  axial, 43
  circumferential, 10, 96, 101
  continuity of, 147, 151
  derivatives of, 27, 108, 131
  field, 42, 43, 151
  finite. *See* large
  formulation, 26, 28, 29–36, 55
  function(s), 29, 58, 84, 102, 135–7
  generalized, 29, 31, 38, 39, 85, 87, 152
  global, 38, 40
  ground, 85
  incremental, 132
  initial, 85
  large, 2
  lateral, 42
  local, 39
  member, 42
  meridional, 10, 99, 100, 171
  method, 63, 67

Displacement(s)—*contd.*
  middle surface, 2, 10, 12
  nodal, 4, 32, 33, 39, 131–5
  nonlinear, 135, 136
  normal, 10, 43, 58, 59, 95, 96, 116–19, 122, 123, 125, 126, 171
  pole, 15
  relative, 86, 88, 116
  rigid body, 19, 20, 33, 35, 36
  rocking, 114
  of shells of revolution, 3
  translation, 114
  variable, 34
  vector, 21, 31, 33, 34, 39, 77, 85, 88, 99, 112, 114
Distributed load vector, 27
Domain(s), 3
Dome, 15, 16
Donnell, L., 91
Doubly-curved elements. *See* Elements, curved; general shell
Dunhamel integral, 85
Dynamic analysis, 77, 84, 95, 166
Dynamic response, 104
Dynamic wind load. *See* Loading, dynamic; wind

Earthquake
  analysis, 89, 95–122
  horizontal, 85, 86, 96, 108, 110, 112, 180
  loading, 115, 116, 120, 179, 180
  motion, 116
  response, 110
  vertical, 85, 86, 112
  see also Seismic loading
EASI, 83
Effective width, 37, 38
Eigenvalue(s), 36, 82, 83, 138, 139, 171
Eigenvector, 82, 106, 109, 138
  interactive, 106
  normalized, 180–3
  see also Mode shapes
Elastic law. *See* Constitutive law
Elastic modulus. *See* Young's modulus
Elasticity, theory of, 2
El-Centro earthquake, 87, 115
Element(s)
  axisymmetric. *See* rotational shell
  bar, 44
  cap, 20, 34, 65
  closed, 15, 30
  compatible, 147, 151
  conical frustum, 3, 66
  core, 102
  curved, 3
  damping matrix, 78
  domain, 24
  equilibrium equations. *See* Equilibrium equations, element
  general shell, 144, 147–55, 158–9
  high order, 84
  high precision, 101, 104, 175
  isoparametric, 101
  low order, 84
  mass matrix. *See* Mass matrix, element

203

Index

Element(s)—contd.
 open-type, 41–3, 46, 47, 49, 52, 104, 172, 175, 181, 183
 quadrilateral, 164
 ring, 8–10
 rotational shell, 8, 35, 48, 56, 123, 147–9, 151–8, 165
 solid, 101
 stiffness matrix. See Stiffness matrix, element
 super-parametric, 149
 transition, 147, 148, 151–155, 158, 159
 variable order, 83–84
Elliptical cap, 38
Elliptical dome, 42
El-Shafee, O. M., xiii, 52, 75, 101, 102, 104, 108, 128
End conditions, 43
 see also Boundary conditions
Energy
 complementary strain, 22–3
 formulation for shells, 3
 functional, 20, 171
 kinetic, 20, 21, 171
 potential, 171
 strain, 23, 171
 total potential, 20, 131, 132
 see also Variational principles
Equations of motion, 86, 124
 condensed, 78
 element, 77, 78
 global, 78–9
 uncondensed, 77
Equilibrium equations
 condensed, 31–2, 132
 element, 29–31, 132
 global, 32–6, 138, 144
 modified, 131–2, 139
 see also specific types of shells
Equilibrium state, 131
Equivalent boundary system(s), 101, 105
Errors, 3, 83–4
Extensional stress. See Stress, extensional

Fagel, L. W., 100, 128
Fast Fourier Transform (FFT), 89–90, 118
Finite differences, 2, 169, 171
Finite element method, xiii, 1, 20
Finite elements. See Elements
Firmin, A., 45
Flexural rigidity, 18, 45
Floegl, H., 143, 146
Fluid-filled cylindrical shell, 123–7
Fonder, G. A., 35, 75
Force(s)
 distributed, 12
 formulation, 26, 29
 in ring footing, 110, 111
 line, 12
 prescribed, 23
 seismically induced, 108
 vector, 89
 see also Stress resultant
Formulation(s)
 displacement. See Displacement formulation

Formulations—contd.
 dynamic, 1
 force. See Force formulation
 mixed. See Mixed formulation
 stability, 1. See also Instability
Forray, M. J., 24, 25
Foundation
 damping, 115
 fixed, 106
 flexibility, 116
 interactive, 100–2
 pile. See Pile foundation
 response, 110–12
 shallow, 100–2
 stiffness, 115, 118
 uplift, 110, 112
Fourier
 coefficient, 19, 31, 54, 125, 133, 148, 156, 179
 components, 14, 131, 161, 179
 harmonic, 101, 165
 series, 4, 12, 13, 49, 130, 147, 178
Framework, circular, 41
Free vibration analysis, 35, 82–5, 96, 104–8, 123–7, 138, 179–81, 190
Frequency
 damped, 89, 120
 dependence, 78, 89, 101, 114
 discrete, 89
 domain, 89–90, 114, 118
 driving, 106, 112
 fundamental, 104, 112, 115
 history, 117, 118
 natural, 79, 82–5, 87, 91–6, 108, 124, 127, 138, 179–81
 predominant, 89
 resonant, 118
Functional, energy, 4

Gallagher, R. H., 24, 25
Gaus, M., xiii
Gauss–Codazzi relations, 7
Gauss point(s), 56
Gaussian curvature, 20, 142
Gaussian elimination, 33, 55
Generalized coordinates. See Coordinates, generalized
Geometric data, 10
Geometric nonlinearity, 130–8
Geometric parameters, 9
Geometric stiffness matrix. See Stiffness matrix, geometric
Geometry, xiii, 1, 170, 174–5
 continuity of, 151
 continuous, 1
 surface, 7, 143
 see also specific types of shells
Geradin, M., 78, 127
Gere, J. M., 45, 75, 187, 199
Global equations of motion, 78
Global equilibrium equations, 32, 88, 89
Global grid, 35. See also Mesh
Global nodal locations, 35
Goldberg, J., xiii
Gould, P. L., 1, 4, 5, 7, 16–18, 20, 24, 25, 27, 33,

Gould, P. L.—*contd.*
  42, 51, 52, 56, 58, 66, 75, 76, 78, 80, 82–5, 90, 96, 97, 99, 101, 102, 104, 108, 110, 112, 114, 115, 123, 124, 128, 129, 132–8, 145, 148, 149, 151, 152, 161, 168, 171, 174, 199
Grafton, P. E., 3, 5
Grief, R., 122, 125, 129
Gravity loading. *See* Loading, self-weight
Ground acceleration, 108
  motion, 85, 86, 89, 112
  see also Earthquake; Seismic loading
Gupta, K. K., 83, 128

h-convergence, 56, 66
Half-space, elastic, 101
Hamilton's Variational Principle (HVP), 20–2, 77
Han, K. J., xiii, 148, 149, 151, 152, 161, 168
Hanna, S. L., 143–5
Harmonic, 12, 79, 140, 141
  analysis, 3, 26
  coupling, 132, 154–7
  dependent, 31
  form, 30
  number, 110
  see also Fourier series
Hemispherical shell
  free vibrational of, 83, 93–5
Hermetian interpolation functions, 42, 80
Hierarchic functions, 27
High-precision formulation. *See* Element, high-precision
Hill, D. W., 3, 5, 66, 76
Historical review, 3–4
Hoop stress. *See* Stress, cricumferential
Horizontal radius, 1–2
Houbolt's method, 90
Hydrostatic loading, 59
Hyperbolic cooling tower, 41, 95–122, 104–12, 116, 139–41, 173
  see also Hyperboloidal shell of revolution
Hyperboloid. *See* Hyperboloidal shell of revolution
Hyperboloidal shell of revolution, 57, 58, 65
  buckling of, 139–41
  column-supported, 173–99
  cornice of, 96, 139–41, 187
  discretization of, 175–77
  dynamic analysis of, 95–100
  earthquake load on, 96–100, 108–12, 174, 179–81, 187–99
  equation of meridian, 107
  free vibration of, 83, 179–81
  geometry of, 140, 174–5
  imperfect, 157–64
  lintel of, 96, 139–41
  mode shapes for, 98, 99, 182, 183
  natural frequencies of, 98, 181, 187
  normal loading on, 57
  on an interactive foundation, 100–22
  pile-supported, 114–22
  ring-stiffened, 190
  static wind loading on, 65–7, 144, 157, 159, 161–4, 178–9, 185–6
  stresses in, 181–5, 188–98

Hyperboloidal shell of revolution—*contd.*
  throat of, 96, 174
  under dynamic wind load, 123, 125, 126
  under self-weight, 157–61, 164, 173, 177, 181–5
  See also Hyperbolic cooling tower

Impedance function(s), 116
Idriss, I. M., 127, 129
Imperfection(s), 141–3, 154–63
  axisymmetric, 159, 164
  bulge type, 157–64
  construction, 141
  fabrication, 142
  sensitivity, 142
Inclined members. *See* Column-supported shells
Incremental nonlinear analysis, 143–5
Initial stress stiffness matrix, 134
Instability, 66–8
  analysis, 130–2, 138–41, 144–5, 171
  ovaling, 187
  see also Buckling
Interactive foundation, 100, 114–22
Intervals, 3
Interelement continuity. *See* Continuity, interelement
Interpolation
  functions, 28, 79, 80, 135–7
  intra-element, 32
  low-order, 55
  matrix, 131
Irons, B. M., 2, 4, 148, 149, 168
Isotropic shell(s), 17–19

Johnson, D. E., 122, 125, 129
Jones, R. E., 3, 5

Kalnins, A., 94, 95, 128
Key, S. W., 164–8
Khojasteh-Bakht, M., 3, 5
Kinematic boundary conditions. *See* Boundary conditions, displacement
Kinematic condensation. *See* Condensation, kinematic
Kinematic conditions, 33
Kinematic constraints, 33
Kinematic law(s), 14–17, 21, 22, 27, 34, 99, 171
Kinematic variable(s), 10–14
Kinetic energy. *See* Energy, kinetic
Kirchhoff's Hypothesis, 2, 14
Klein, S., 3, 5, 130, 145
Krätzig, W. B., 190, 199
Kraus, H., 94, 95, 128

Lagrange multipliers, 171
Lagrangian polynomials, 9
Lambe, T. W., 127, 129
Laplace's equation, 124
Large deflections. *See* Displacements, large
Layered foundation, 101, 114, 115
Layered shell, 18
Layering, 1
Lee, B. J., xiii, 112, 114, 115
Length of shell meridian. *See* Arc length
L'Hopital's rule, 10
Lift(s), 177

Index

Limit point analysis, 143, 144
Lin, J. S., xiii
Line loading. *See* Loading, line
Linear shell equations, 3
Liu, S. C., 100, 128
Lintel, 96, 101, 139, 141
Load vector, consistent, 79
Load(s), Loading, xiii, 1, 170, 171, 174
   axisymmetrical, 169, 171
   antisymmetrical, 64
   applied, 13
   blast, 122, 124
   circumferential, 4, 55
   concentrated, 14, 74
   consistent. *See* Consistent load vectors
   dead. *See* self-weight
   distributed, 13, 14, 186
   dynamic, 1, 139
   earthquake. *See* Response spectrum
   edge, 52, 56, 58, 59
   generalized, 85, 87
   gravity. *See* self-weight
   hydrostatic, 59
   incremental, 132
   inertial, 12
   internal pressure, 66
   line, 4, 55, 110, 171
   mechanical, 12
   non-axisymmetrical, 4
   nonsymmetrical, 1, 66, 169
   pressure, 66, 73, 74. *See also* Pressure
   radial, 64
   ring, 12, 13, 19, 51, 52
   self-weight, 51, 141, 157–61, 164, 173
   self-equilibrated, 52–5
   smooth, 1
   static, 1, 139
   surface, 12, 13, 52, 54, 171
   tensile, 164–7
   thermal, 1, 12, 51, 68, 77
   vector(s), 31, 51–2, 85
   wind, 65, 66, 123, 125, 139, 144, 157, 161–4, 173, 178, 185
   see also specific types of loads
Local effects of columns on shell, 52–5
Local irregularities. *See* Imperfections
Locally nonaxisymmetric shells, 147–68
Lowrey, R. D., xiii
Lu, Z. A., 3, 5
Lukasiewicz, S., 14, 25

Mang, H. A., 143, 146
Mass, xiii
   consistent, 98
   density, 21, 91, 93
   fluid, 127
   structural, 127
Mass matrix, 27, 84, 138
   added, 124
   consistent, 79–82, 96, 172
   effective, 82
   element, 77, 82
   generalized, 85
   global, 79

Mass matrix—*contd.*
   member, 80, 81
Material(s)
   composite, 14
   laminated, 14
   layered. *See* Layering
   law. *See* Constitutive law
   properties, 17
Matrix
   interpolation, 29, 30, 131, 138
   member, 46
   operator, 15
   symbols for, 4
   see also specific types
Maximum–minimum problem, 24
Mebane, P. M., 35, 75
Membrane
   force, 3
   strain energy, 22
   theory, 131, 138, 165
Meridian, 7, 107
   length of, 8. *See also* Arc length
   shell, 32
Meridional angle, 7, 8
Meridional displacement. *See* Displacement, meridional
Meridional discontinuity. *See* Discontinuity, slope
Meridional force. *See* Stress resultant, meridional
Meridional mode shape. *See* Mode shape, meridional
Meridional moment. *See* Stress couple, meridional
Meridional rotation. *See* Rotation, meridional
Meridional strain. *See* Strain, meridional
Meridional transverse shear strain. *See* Strain, transverse shearing
Meridional variable, 7, 8, 26
Mesh, 148, 151, 154, 158, 164–6
Microcomputer program, xiii, 174, 175
Middle surface, 7, 12
Mixed equilibrium equations, 55
Mixed formulation, 22, 26, 28, 55, 58. *See also* Variational principle(s), Reissner's
Mixed matrix, 55
Mixed method, 60, 61, 63, 64, 66, 67
Modal superposition solutions, 82–90
Mode(s), Mode shapes, 84, 85, 87, 91, 95–101, 115, 138, 140, 142, 143, 180–3
   axial, 127
   meridional, 79, 82, 88, 106–9
Modeling, xiii
Modulus of elasticity. *See* Young's modulus
Mokhtarian, K., 70, 76
Moment(s)
   banding. *See* Stress couple(s)
   distributed, 12
   equilibrium, 22, 23
   gradient, 14
   line, 12
   of inertia, 80
   overturning, 186
   resultant(s). *See* Stress couple(s)
   twisting. *See* Stress couple, twisting
   vertical, 110–11
Momentless state of stress. *See* Membrane theory

Index

Motion
  equations of, 78–9
  rigid body. *See* Displacements, rigid-body
Multilayered. *See* Layered foundation; Layered shell

Navaratna, D. R., 3, 4, 5, 6, 131, 132, 139, 145
Natural frequency. *See* Frequency, natural
Newmark's-β method, 90
Newton–Raphson iteration, 144, 171
Nguyen, B., 190, 199
Nodal displacements. *See* Displacements, nodal
Nodal circle(s), 3, 8, 35, 147, 148, 152–7, 175
Nodal force
  consistent, 4
  equivalent, 55
  force vector, 32, 40, 42
  point(s), 8, 147, 153
  variable(s), 4, 29, 43, 55, 133
Nodes
  internal, 153
  line, 147, 148
  moving, 151
  point, 148–57
  sub-, 151, 152
Nonlinear analysis, 130–45, 166, 171
Nonlinear kinematic law, 16, 17, 130–8
Nonlinear material, 2, 131
Nonlinear prebuckling state. *See* Prebuckling state
Non-axisymmetric shells, 147–9, 154
Nonsymmetric loading. *See* Loading, nonsymmetric
Normal displacement. *See* Displacement, normal
Normal vector, 149
Notation for matrices, 4
Novozhilov, V., 20, 25, 130, 145
Numerical integration, 3, 91, 122, 169
Numerical solutions, xiii, 1, 3

Obrecht, H., 143, 146
Open shells, 36
Open-type elements, 41–52, 80–2
Orthogonal curvilinear coordinates. *See* Coordinates, curvilinear
Orthogonality, 84
Orthotropic shell, 17, 18, 19

p-convergence, 56, 68, 66
Pandya, V., 104, 128
parabolic shell, 64–6
paraboloid of revolution. *See* Parabolic shell
Parallel circle. *See* Nodal circle
parallel plane(s), 8
parallel surface, 20, 21, 77
Parameters, definitive, 10
Particular solution, 84
Penzien, J., 3, 5, 84, 89, 128, 179, 199
Percy, J. H., 3, 5
Period, 87
  fundamental, 123, 181
  of vibration, 87
  see also Frequency, natural
Periodic function. *See* Fourier series
Perturbation of equilibrium state, 131, 132

Peters, H. L., 190, 199
Phan, L. T., xiii, 124, 129
Phase angle, 89
Pian, T. H. H., 3–5, 131, 132, 139, 145
Pile foundation, 114–16
Pipe connections, 147
Plane(s), parallel, 8
Plastic buckling analysis, 143
Point load. *See* Load, concentrated
Points
  scaled, 10
  tabulated, 10
Poisson's ratio, 18, 37, 91, 93
Polar angle, 4
Polar moment of inertia. *See* Moment of inertia
Pole of a shell, 8
Polynomial(s), 27
  approximation, 4, 83, 135
  cubic, 35
  first order, 135–7
  Hermetian. *See* Hermetian interpolation functions
  high(er) order, 4, 62, 132, 138
  interpolation, 10, 92, 95
  Lagrangian. *See* Lagrangian polynomials
  low order, 84, 135, 136
  second order, 135
  shape function. *See* approximation, interpolation; Shape function
Popov, E. P., 3, 5
Position vector, 149
Post-buckling analysis, 143
Potential energy. *See* Energy, potential
Prato, C. A., 23, 25
Prebuckling
  deformations, 144
  state, 131
  stress matrix, 134
Pressure
  axisymmetric, 140–1
  constant, 123, 139, 140
  critical, 139–41
  distribution, 65, 139, 141, 178–9
  external. *See* wind
  hydrodynamic, 124, 127
  internal, 178. *See also* suction
  stagnation, 145
  suction, 144, 145
  uniform, 161
  velocity, 178, 179
  wind, 123, 139
  see also Load(s), Loading
Pressure vessels, 66
Principal radius of curvature, 7, 174
Principles, energy. *See* Variational principles

Radial loading, 64
Radius of curvature, principal. *See* Principal radius of curvature
Rayleigh–Ritz method, 24
Reactions, discrete, 52–5. *See also* Column(s)
Reduced integration, 149
'Reinforced Concrete Cooling Tower Shells', ACI Title, 74, 179, 199

Index

Reinforcing steel, 186
Reissner, E., 22, 25
Reissner–Naghdi theory, 95
Reissner's Variational Principle (R.V.P.), 22–4, 27
Rensch, H. T., 143, 146
Resonant frequency. *See* Frequency, resonant
Response spectrum, 87–8, 110, 112, 180, 187
   analysis, 96–100, 108–12, 114, 119, 172, 179–81, 187–99
   maximum, 119, 121
   pseudo-velocity, 87–8
   velocity, 97, 99
Reticulated shell. *See* Shell, reticulated
Rigid-body displacements. *See* Displacements, rigid-body
Ring(s)
   axially compressed, 144
   beam, 97, 105, 187, 190
   element. *See* Element, ring
   footing, 101, 109, 110
   load(s), 30, 31, 58
   load vector, element, 27
   stiffener(s). *See* Stiffeners, ring
Robinson, A. R., 91, 92, 112, 128, 129
Rock foundation, 101, 102, 104–6
Rocking, 110–12, 118
Root-sum-square (RSS), 88
Rosemeier, G., 190, 199
Rotational shells. *See* Shells of revolution
Rotation(s)
   about normal, 48, 49, 81
   circumferential, 10
   meridional, 10, 171
   moderate, 16
   nonlinear, 131
   of coordinate axis. *See* Coordinate transformation
   prescribed, 23
Rotational shells. *See* Shells of revolution
Rotatory inertia, 95
ROT B, 13, 70
Rotter, M., 74, 76

Scope of treatment, 1–2
Schallert, K., xiii
Schnobrich, W. C., 91, 92, 128
Seed, H. B., 127, 129
Seide, P., 143, 145
Seismic loading, 64, 100, 120. *See also* Earthquake; Response spectrum
Self-equilibrated edge effects, 52, 109
Self-equilibrated loads, 110
Self-weight, 65, 110
Sen, S. K., xiii, 20, 25, 56, 58, 75, 78, 83–5, 96, 99, 128, 129
Setlur, A. Y., 104, 128
Shape factor, 59, 62
Shape function(s), 26, 27, 147, 148, 151. *See also* Comparison function(s); Polynomial(s)
Shear(ing)
   factor, 18
   lag, 38
   modulus, 18, 107
   rigidity, 18

Shearing—*contd.*
   strain. *See* Strain, shearing; Transverse shearing
Shell-like frames. *See* Shells, reticulated
Shell(s), 1
   classical analysis of, 14
   closed, 7, 8, 20, 33, 38
   complex, 41
   deformations of. *See* Displacement(s)
   doubly-curved, 3
   displacements of. *See* Displacement(s)
   element. *See* Element(s)
   equilibrium of. *See* Equilibrium equations
   geometry of. *See* Geometry; specific types of shells
   imperfect. *See* Imperfection(s)
   instability of. *See* Buckling; Instability
   linear theory of, 15
   multisegment, 107
   of revolution, 1–3, 10, 24, 26, 41, 132
   on columns. *See* Column-supported shells
   reinforcement, 108
   reticulated, 18
   theory, classical, 2, 49
   theory, conventional, 41
   thickness of. *See* Thickness
   thin, 1
   wall construction, 170
   see also specific types of shells
SHORE, xiii, 2, 59, 62, 66, 71–3, 90–2, 94–5, 122, 123, 125, 126, 139, 169, 171–5, 179
Shye, K. Y., 112, 129
Sign convention, 11
Simple support. *See* Boundary, hinged
Single-degree-of-freedom system (SDF), 87
Slope, 3
Soil
   conditions, 104, 106
   density, 115, 127
   depth, 107
   friction, 112
   layered, 105
   medium, 101, 105, 107
   model, 106
   properties of, 127
   reactions, 101
   stiffness. *See* Foundation stiffness
Soil–pile–structure interaction, 114, 118
Soil–structure interaction, 89, 100–22
Solution(s)
   analytic, 1, 2
   numerical, 1
Space Frame, 45
Spectrum. *See* Response spectrum
Spherical shell
   cap of, 38, 66
   closed. *See* Dome
   column-supported, 52
   pressurized, 33
   see also Hemispherical shell
Spierig, S., 190, 199
Split rigidity, 37
Spring
   foundation, 101, 102
   static, 110

Index

Spring—*contd.*
  stiffness, 102, 104
Stability. *See also* Buckling instability analysis, 16, 166
Static analysis, 26–76
Static condensation. *See* Condensation, static correction, 101, 102, 183, 186
Statical variable(s), 10–14
Stations, 3
Stationary problem
  global, 24
  local, 24
Stationary Total Potential Energy, Principle of (PSPE), 22, 28, 29
Stein, E., 190, 199
Steinmetz, R. L., 123, 126, 129
Stiffened shell, 18, 36–8, 41, 79, 80, 172
Stiffener(s)
  area of, 37
  channel, 36
  circumferential, 36, 190
  eccentric, 36
  H, 36
  meridional, 36, 38
  moment of inertia of, 37
  multiple, 38
  plate, 36, 37
  ring, 36, 141, 169, 171
  spacing, 38
  stringer, 36
  symmetrical, 38
Stiffening, 1
  circumferential, 141, 190
  rib. *See* Stiffener(s), meridional
  ring. *See* Ring beam
Stiffness matrix, 27, 84, 89
  branch, 41
  composite, 144
  displacement, 144
  effective, 49–51
  element, 31, 32, 50, 77
  generalized, 85, 87
  geometric, 132–8, 144
  global, 33, 40, 41, 79, 138, 148, 152–7
  harmonic, 156
  initial stress, 134
  linear, 144
  local, 45
  member, 43–5, 46, 48, 49
  pressure-rotation, 144
  reduced, 48–9
Stiffness method, direct. *See* Direct stiffness method
Strain(s), 6–17, 22, 27, 34, 35
  circumferential, 11
  energy density function, 27
  energy intensity, 57, 58
  extensional, 11
  in-plane, 11, 130
  meridional, 11
  middle surface, 11, 13
  nonlinear, 16–17, 131–3
  transverse shearing, 2, 10, 14, 17, 27, 44, 59–62, 66, 130, 135, 136, 149

Strain(s)—*contd.*
  warping, 44
Strain–displacement relations, 16, 43, 108, 130, 132–8
Stress(es), 27, 35
  analysis, 1
  bending, 165
  calculation of, 34, 35, 84, 88, 92, 99, 171, 187
  compressive, 186
  circumferential, 72, 73, 183, 184
  couple(s), 11, 13, 17, 22, 35, 38, 40, 53, 55–8, 108, 118, 122, 134, 137, 161–4, 174, 183, 186, 190, 193, 195
    circumferential, 11, 108, 137, 161–4, 183, 186, 191, 193, 195
    meridional, 11, 58–61, 67, 68, 108, 120, 134, 137, 161–4, 183, 184, 186, 190, 193, 195
    twisting, 11
  extensional, 70, 165
  initial, 132
  intensity, 166–7
  prebuckling, 17, 134, 138, 139, 171
  meridional, 71, 74
  normal, 149
  resultant–displacement relationship. *See* Constitutive law
  resultant(s), 11, 13, 17, 22, 34, 35, 38, 40, 53, 56, 62, 99, 103, 108, 112, 118, 120–2, 134, 137, 139, 159–64, 174, 183–6, 188, 189, 192, 194
    circumferential, 11, 68, 103, 108, 119–23, 134, 137, 139, 159–64, 171, 181, 184–6, 188, 189, 192, 194
    hoop. *See* circumferential
    in-plane shearing, 11, 186, 189, 193
    meridional, 11, 65, 68, 103, 108, 120–3, 126, 134, 137, 139, 159–64, 171, 181, 184–6, 188, 192, 194
    transverse shearing, 12, 14. *See also* Transverse shearing force
  vector, thermal, 17
Stricklin, J. A., 3, 5, 27, 35, 75, 130, 131, 135, 145
Stringers, 36, 38
Strome, D. R., 3, 5
Structural form, description of, 1
Sturm sequence method, 83
Subdomain, 24
Substructure, 154–7
Superposition, 101
Surface(s)
  area, 31
  bounding, 2
  geometry, 7
  loads. *See* Loads, surface
  middle, 2, 8, 18, 77
  reference, 2, 21
  variable(s), 31
Suryoutomo, H. B., xiii, 4, 5, 85, 96, 99, 128
Symmetry, 33
Szabo, B. A., 4, 5, 56, 75

Thermal coefficient, 19
Thermal gradient, 141
Thermal load. *See* Load(s), Loading, thermal

Index

Thermal load vector, element, 31
Thickness, 1, 108, 149, 177, 187
  basic, 37
  bending, 19
  effective, 18
  equivalent, 37, 38
  extensional, 18
  membrane, 37
  profile, 140–1
  variable, 10
  variations in, 139
Time
  dependence, 88
  domain, 118
  history, 90, 119–20, 123, 125, 126
  integration, 21, 90
  step, 90, 122
  variables, 10, 12
Timoshenko, S., 58, 62, 66, 75, 187, 199
Too, J. M., 149, 168
Torospherical head, 66
Torospherical geometry, 68
Torospherical knuckle, 66
Torsion
  shell, 14
  surface, 11
Torsional rigidity, 45
Transfer matrices, 4, 169
Transformation
  matrix, 133, 152
  vector, 113, 114
Transitional surfaces, 38
Translation, 110–12, 118
Transverse shearing
  force, 58–64
  strain. *See* Strain, transverse shearing
  stress. *See* Stress, transverse shearing
Trappel, F., 143, 146
Trial function. *See* Comparison function
Trojan tower shell, 140, 142, 144
Twist of surface, 11
Twisting
  moment. *See* Stress couple, twisting
  rigidity, 18

Ultimate strength analysis, 143
Uniform base motion. *See* Base motion
Unit
  vector, 150
  weight, 177

Unsymmetrical loading. *See* Loading, nonsymmetrical
Uplift, 102, 106

Van Dyke, P., 164–8
Variable(s)
  condensed, 32, 34
  dependent, 10, 22, 24, 26
  intra-element, 31
  kinematic, 10
  nodal, 131
  separation of, 12
  static, 10
Variational principle(s), 20–4
  Hamilton's, 20–4, 77
  Reissner's, 22–4, 27
  Stationary Total Potential Energy, 22
Variational problem, solutions of, 24
Vector. *See* specific types of vector
Velocity
  ground, 85
  initial, 85
  instantaneous, 77
  pressure. *See* Pressure, velocity
Viscoelastic material, 114

Walter, H., 143, 146
Warping strain, 44
Wave(s)
  axial, 127
  circumferential, 140. *See also* Harmonic
  meridional. *See* Mode shapes
  number. *See* Harmonic number
Weaver, W., Jr., 45, 75
Weight of a shell. *See* Self-weight
Weighting
  factor, 24
  function, 9
Weingarten, V. I., 91–3, 128
Whitman, R. V., 127, 129
Wilson-θ method, 90
Wind. *See* Loading; Pressure
Witmer, E. A., 66, 75, 131, 132, 139, 145
Woinowsky-Kreiger, S., 58, 62, 66, 75
Wunderlich, W., 70, 76, 143, 146

Young's modulus, 18, 91, 93
Yu, W. W., 38, 75

Zerna, W., 190, 199
Zienkiewicz, O. C., 2–5, 132, 145, 148, 149, 168